W9-CJB-386

Award-Winning Passive Solar Designs

Professional Edition

Jeffrey Cook

McGraw-Hill Book Company

New York St. Louis San Francisco Auckland
Bogotá Hamburg Johannesburg London Madrid
Mexico Montreal New Delhi Panama Paris
São Paulo Singapore Sydney Tokyo Toronto

1234567890 89876543

ISBN 0-07-012478-7

Library of Congress Cataloging in Publication Data

Main entry under title:

Award-winning passive solar designs.

"Selections—award winners and finalists—from the First Passive Solar Design Awards Competition"—Pref.

1. Solar houses—Design and construction—Competitions. 2. Solar energy—Passive systems. I. Cook, Jeffrey.
TH7414.A92 1984 728'.69 83-19551
ISBN 0-07-012478-7

Credits

Illustrations
Michael W. Jager, pages 40–41, 52–53, 76–77, 82–83, 94–95, 108–109, 132–133, 144–145, 150–151, 194–195, 222–223, 274–275.

Photos
Norman McGrath, pages 20, 21.
C. Carey, page 28.
Conde Nast Publications, pages 34, 39.
Alternate Design, page 53.
Mark Citret, page 53.
Darrow M. Watt, page 138.
Doug Lee, pages 168, 175.
Stuart White, page 184.
Joseph W. Molitor, pages 206, 210, 211, 212, 213.
The Colyer/Freeman Group, page 214.
Franz Hall, pages 228, 233.
O. Baitz, Inc., pages 236, 240.
Jerry Goffe, pages 242, 246, 247.
ELS Design Group, pages 258, 262, 263.
Esto Photographics Inc., pages 266, 270, 271, 273.
Photography Unlimited, page 280.

Contents

Preface

This book illustrates some of the best passive solar house designs available at the beginning of the 1980s, including 15 projects added specifically for the professional market. These are selections—award winners and finalists—from the First Passive Solar Design Awards Competition. This international competition attracted more than 350 entries from the United States, Canada, Europe and several Asian countries. A distinguished jury of architects, engineers and other professionals judged these designs for their "excellence in the synthesis of architecture and engineering."

The quantity of first-class entries was almost startling. Included were modest shelters and expensive country homes. Although every entry had the same submission requirements, including thermal performance calculations and technical summaries, each was judged on its own merits. A design with a high solar fraction did not automatically become an award winner. Neither did an exceptional architectural design with incomplete thermal analyses.

In many cases, the technical advisors to the jury performed independent thermal calculations on minicomputers to confirm the validity of information provided by designers. Inevitably, in such a large competition, there were superior designs that did not get selected. But, in the view of the jurors, each that was recognized achieved an exemplary synthesis of solar architecture and engineering. The distinguished jurors were:

Dr. J. Douglas Balcomb, research scientist, Los Alamos Scientific Laboratories, Los Alamos, New Mexico and chairman, American Solar Energy Society, United States Section/International Solar Energy Society.

Peter Calthorp, architect, VanDerRyn, Calthorp and Partners, Inverness and San Francisco, California.

William Caudill, architect, Caudill, Rowlett, and Scott, Houston, Texas.

Ralph Johnson, research engineer, president, National Association of Home Builders Research Foundation, Rockville, Maryland.

Douglas Kelbaugh, architect and solar consultant, Kelbaugh & Lee, Princeton, New Jersey.

William M. C. Lam, lighting engineer, William Lam Associates, Cambridge, Massachusetts.

Richard Rush, engineer and senior editor, *Progressive Architecture*, Stamford, Connecticut.

In this book, both prizewinners and finalists are presented and organized according to themes based on climate, construction or building type. The discussion and information in the text is based upon the designers' competition portfolios, as well as additional materials provided since the competition was held in October 1980 at Amherst, Massachusetts. In some cases, unbuilt designs have been subsequently constructed, sometimes identical to the preliminary design, in other cases with modifications.

The sponsor of the First Passive Solar Design Awards Competition was the Passive Systems Division of the American Solar Energy Society; the organizer and coordinator was the New England Solar Energy Association. Subsequent to the competition, many of these award-winning designs have been honored in other competitions and award programs. Many have appeared in prominent magazines, periodicals and books. In a sense, all these dynamic events have made the creation of this book more difficult, but also more important: They confirm the quality of these solar homes and the validity of solar building as a critical stream in contemporary architectural theory and practice.

Jeffrey Cook

Introduction

There is something special about associating solar energy with the design of a home. Perhaps part of the attraction that links the two is a common idealism: solar energy is the ultimate energy source on our planet, and thus represents the ideal generator of any thermal or mechanical advantage. Similarly, the ideal and most desired residential type in the United States, even for apartment dwellers, is typically the single-family house. As a dwelling type, the house offers the most options for residential accommodation. Therefore, it should not be surprising that the first applications of solar energy to buildings have been to single-family houses.

More important: in developing designs that respond to the sun—the fundamental driver of the earth's climate and energy systems—architects are reconceptualizing the theoretical basis of building design. More simply put, solar design signifies a rebirth of design theory, a return to the fundamentals of nature, with profound architectural implications.

In some ways, this book explores and tries to capture this renaissance. It presents the work of creative, imaginative designers who have made their return to the fundamentals of nature, who have understood the sun's energy and who have devised ingenious and attractive ways to use it in their home designs. However, it is difficult, if not foolhardy, to attempt any sweeping generalities or summaries of these designs. Certainly, they represent an enormous diversity of housing solutions. In fact, the differences are so various, it is difficult to identify many common themes.

Any assumption that strong energy requirements by themselves bring automatic design conformity is immediately disproved by these creative designers. If anything, the discipline of solar design has stimulated the goal of design individuality within a common theme of solar dependence. The assumption about solar houses that the universality of the sun and the limited geometry of its sky path would force some ideal solar profile on house design has proved to be wrong. Similarly, there is no favorite or preferred method of passive heating or cooling.

Passive Solar House Design

What, exactly, is a passive solar house design? This popular term has been used since about 1975 to describe houses heated by the sun without the use of mechanical equipment. The term "passive" is also used to describe certain natural cooling processes that allow heat to dissipate from buildings to the surrounding environment. In both cases, human comfort is the first goal. In passively-heated homes, the whole house becomes a solar-heat collector, rather than some piece of hardware mounted on the roof. Passive designs attempt to manage natural thermal flow through a house by the shape, composition, orientation and management of various construction assemblies.

In contrast, "active" solar systems use pumps and fans powered by secondary sources to collect, distribute and manage the sun's free energy. But, with both active and passive solar applications, the use of some back up system to deliver comfort when the weather does not cooperate is the practice. Designing the appropriate mix of passive systems and conventional back up systems requires a special design skill. And a design competition can be an important form of encouragement for those who are so skilled.

Whether a solar design is "passive" or "active," its success is not necessarily related to the efficiency of the design, the ratio of "energy in" versus "energy out," or the "purity" of the solar system. Rather, the important question of design quality is the energy return for investment, or effectiveness. Especially in residential applications for space heating, passive solar designs are almost always more economical than active designs or systems.

Other Lessons

Several other lessons about passive solar designs may be derived or at least inferred from the competition that resulted in the selection of these homes. For example, it has been assumed that the critical ratio between the area of solar-collection glazing and the necessary exposed surface of thermal storage mass would force the shape and materials of residential interiors into predictable patterns. But, the increased awareness of the importance of thermal conservation, as represented in insulation standards and displayed here, has itself brought a great diversity to the optimum area of solar heat collecting glazing. There is no single best passive solar solution, even for similar houses in the same climate. And the growing range of thermal storage materials and their strategies of application have increased the design choices even when solar apertures are similar. Thus, even the rigid glass and mass formulas recommended for balanced thermal designs have not resulted in formula house designs.

On the other hand, sensitivity to human comfort needs within a specific climate can reinforce certain common design approaches, and thus encourage regionalism in architectural design. Houses for New England tend to share both bioclimatic and cultural characteristics in comparison to designs for the southeast and southwest. And, although regionally appropriate patterns can be identified, there seems to be no such thing as a single solar style. Passive design is an ethical attitude and a design discipline that can operate within a variety of architectural styles. Certainly not all of these prizewinning house designs will be equally attractive. Fortunately, solar design can work within a broad range of architectural tastes.

Cooling

A troublesome technical question that also has important design implications is cooling. In much of the United States, space cooling in the summer is an important building consideration. Passive cooling has sometimes been conceived as the complement of passive solar heating. But, by definition, passive cooling is "non-solar" or "anti-solar." Unfortunately, passive cooling is much more difficult to achieve than passive solar heating. Passive cooling typically needs some mechanical assistance to improve its effectiveness. In comparison to passive solar heating, such "hybrid" cooling technologies have not been well developed yet, either in terms of performance estimation or in terms of architectural application.

Only two house designs are included here from significantly overheated climates. But, both sunbelt examples, for the warm, damp climate of Savannah, Georgia and the hot, dry climate of Yuma, Arizona, are prototypes in bioclimatic response and in sound architectural detailing for passive cooling. The clarity of these distinct designs emphasizes how different passive design is when all the bioclimatic aspects of different locations are respected.

Models for Many Places

Most of the climate types of the world can be found throughout the North American continent. Thus, to some degree, these passive solar house designs represent many of the appropriate issues and responses for other countries as well. From a cost standpoint, all of these houses were practical within the context of the 1980 economy. So, they are even more practical design models as the years go by. Each year, the annual energy estimates and the bases for resultant design decisions change with general inflation. However, once built, a solar design continues to provide a major share of its energy needs at no cost. The impact of continuous inflation or of lower effective income can only heighten the value of each of these designs.

For the immediate future, it can be seen that passive solar houses are not only for the rich or eccentric. They can be built within the budget of any conventional house or for modest increases in cost against the minimal budget. And, they can dramatically alter the fuel dependence of a dwelling, regardless of climate. Examples on the following pages range from tract houses to custom-designed homes, from speculative condominiums to retrofitted, century-old farmhouses.

Little has been said about some of the more subtle intangibles of passive solar houses. The reigning quiet, the freedom from internal, mechanically-generated building noise, as well as from outside sound, is one of the memorable aesthetic qualities of passive solar homes. Simultaneously, independence from fossil fuel brings dependence on the immediate natural world. The vagaries of the daily weather reinforce our understanding of our relationships both with the immediate environment of nature and with the larger sense of cosmos that is most easily symbolized by our nearest star, the sun. These are among the more personal and more profound reasons for passive solar designs.

A Solid "Solar Woodbox"

RESIDENCE FOR

Shirley and Hugh Kirley
Amherst, Massachusetts

DESIGNER

Sawmill River Post and Beam, Inc.
Leverett, Massachusetts 01054

BUILDER

Shirley and Hugh Kirley
5 Wildwood Lane
Amherst, Massachusetts 01002

STATUS

Occupied, July 1980

Estimated Building Energy Performance

BUILDING DATA

Heating degree days	6,850 DD/yr.
Floor area (each)	624 sq. ft.
Total floor area	1,953 sq. ft.
Building cost	$72,000
Cost of special energy features	$ 7,150
Total building UA (day)	434 Btu/°F. hr.
Total building UA (night)	161 Btu/°F. hr.
Heating	
Area of solar-heat collection	413 sq. ft.
Solar heat	54 percent
Auxiliary heat (wood)	46 percent
Cooling	
Heat avoidance	insulated window shades
Natural means	cross-ventilation
Auxiliary cooling	vent fan 200 hr./yr.
Cost of auxiliary cooling	1¢/hr./sq. ft./yr.
Hot water	
Need	25,555 gallons/yr. @ 140°F.
Auxiliary electricity cost	$35/yr.

ANNUAL ENERGY SUMMARY

Nature's contributions	
Heating	42.0×10^6 Btu
Cooling	no estimate
Hot water	17.7×10^6 Btu
Auxiliary purchased	
Electricity	6.2×10^6 Btu
Wood	less than a cord or 19,964 Btu/sq. ft.

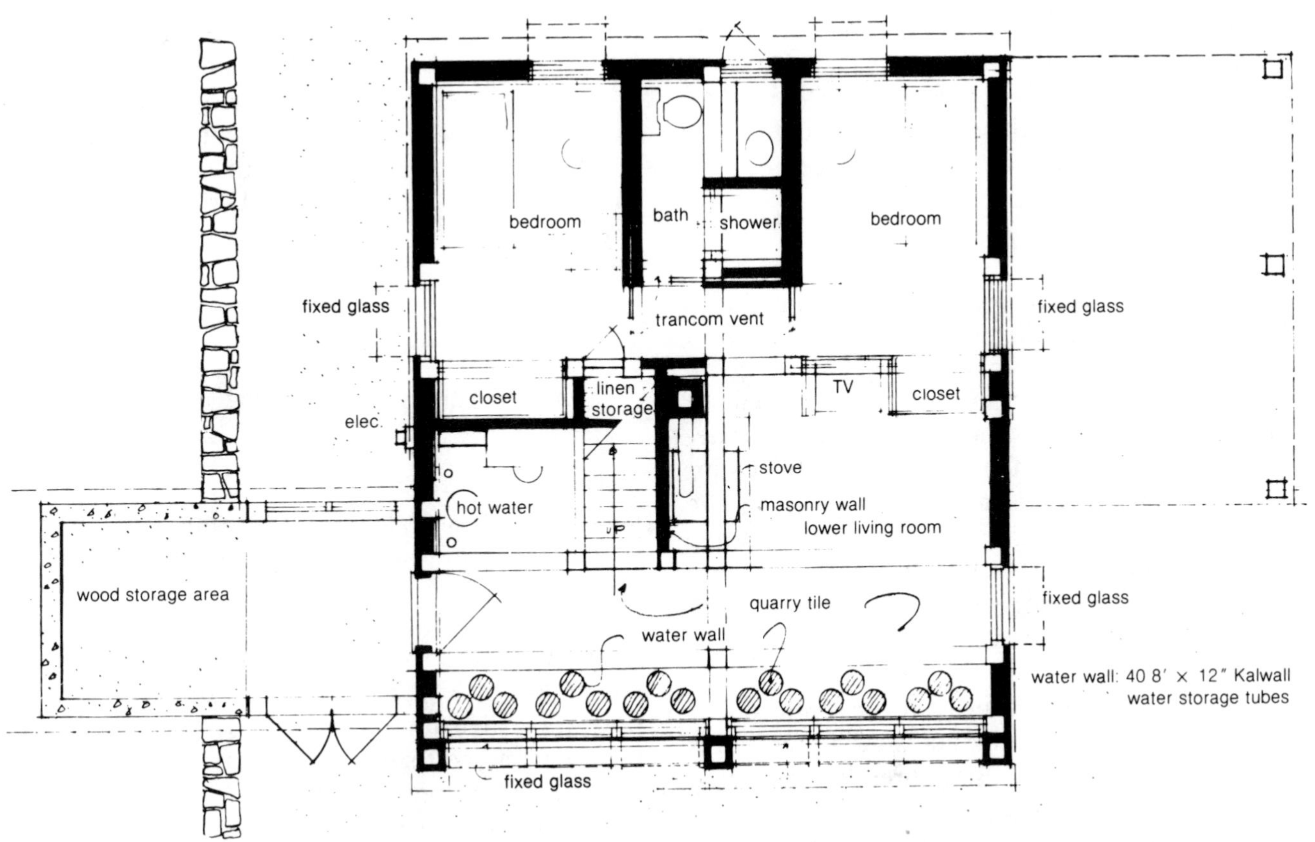

Kalwall water tubes dominate south wall of this floor plan.

In this custom-built solar home, the owners were not trying to invent anything new. Rather, Shirley and Hugh Kirley said they wanted to combine "the best of what we know about our daily living needs, our environment both inside and out and today's technology in post and beam construction, insulation, heating and thermal storage." The result was a nearly cube-shaped, post and beam home they called a "Solar Woodbox."

Shirley is an interior designer and Hugh is a landscape architect and home designer. Together, they own a successful timber-frame home designing and marketing company known as Saw Mill Post and Beam. They designed their home for their family of four, and set a budget of less than $85,000, including appliances, solar system, carport and site work.

They also agreed to build no more than 2,000 square feet. "This helped us develop the basic philosophy of conservation and efficiency in our use of space and energy, which is essential to make a passive solar house work," Hugh Kirley said.

"All of the basic knowledge we have about siting, double entries, compact-open floor plans, central chimneys, wood backup heat, inside vented dryers and stoves, projecting overhangs and concrete mass were combined with the best of today's technology in insulation, caulking, glazing, insulating night shades and passive solar-heat storage.

"Although we used several publications as references, our text book was Edward Mazria's *The Passive Solar Energy Book*.* Our water thermal

* Edward Mazria, *The Passive Solar Energy Book* (Emmaus, Pa.: Rodale Press, 1979).

storage wall was chosen and sized using his criteria, 'to provide enough thermal energy on an average sunny day in January to supply a space with all of the heating needs for the day . . . maintaining an average temperature of 65 to 75 degrees over the period.' "

An Efficient Solar Woodbox

The nearly cube shape of the house provides a maximum of interior volume with a minimum of exterior surface. Such geometry tends to minimize winter heat loss, an important consideration in the cold New England climate. Each of the three floors, including the top-story master suite, is approximately 24 × 28 feet. This energy-efficient form has the complete south wall as a passive collector area.

In addition, the tilt angle of the south roof is ideal for a year-round solar hot water system. This is a closed-loop system, using three roof-mounted *Sunworks* panels, with an 80-gallon hotwater tank and an electrical backup. The delivered-water temperature is set at 140°F.

"Designing" the Microclimate

The Kirleys placed the house on the north side of their property to take best advantage of the sun's rays to the south. Built on a sunken terrace with an existing belt of protective trees to the north and west, the house sits in an advantageous microclimate. Around the house, a "lawn" of washed white stone adds reflected radiation to the solar potential. The carport on the west is not far from the street, but also close enough to the house to help protect it from winter winds. A passive solar link from the carport to the house provides a solar-heated airlock entry and storage.

The main floor, the second level of the house, is totally open without any enclosed spaces. Stairs go up to the master bedroom loft or down to a lower, informal living room, next to the two children's bedrooms. A single central-chimney flue serves the wood-burning stove used as a backup heating system. Both wall vents and an open stair allow the stove-warmed air to rise and circulate through the house. Firewood is stored directly under the entry /link so that the stove can be loaded easily regardless of the weather. Less than a cord of wood is used each winter, and ashes need only be removed once or twice a year.

Functional and Thermal Efficiency

This house has functional, as well as, thermal efficiencies. For example, paths are short and direct. Although small, rooms are neatly planned to allow maximum use. For instance, the children's bedrooms are only 8 × 11 feet, plus a 4- or 6-foot closet. But, each room has two windows and space for an 8-foot length of desk and dresser. Pocket doors that slide into the wall are part of the careful planning and craftsmanship that give this home its economy of space.

Perhaps the most impressive space planning is on the second level, the center of living, dining, kitchen and laundry activities. The complete north wall has 5-foot, 4-inch high, built-in storage with plain white doors in contrast to the exposed pine. The top of the storage wall is used to display plants and pottery beside the continuous strip of north window. This storage wall is also part of the air circulation system: ducts and baseboard heating units are built in.

The dining area has a 5-foot wide sliding glass door that opens to a 12- × 29-foot deck. The deck receives cool summer breezes on three sides, but its location does not affect the collection of solar heat or the insulation quality of the walls during the winter. In addition, it protects a work/play space which is out of view underneath.

The kitchen side of the main floor has work counters, cabinets and appliances along the west wall, with the laundry opposite the stair. The north storage wall is also part of the kitchen. The dishwasher, kitchen sink and microwave oven are part of the work island/breakfast bar that opens up the work area to the rest of the house.

Pattern of Kalwall water tubes permits homeowner to raise and lower the insulated Energy Saver Shades.

A Wall of Water

The most distinctive visual elements of the house are the 8-foot-high, translucent *Kalwall* water tubes inside the south solar glass wall. This "water wall" for thermal storage is more of a three-dimensional, spatial arrangement of varying light and color densities than just a row of 12-inch diameter cylinders. By arranging the 40 tubes in overlapping linear patterns, the windows are accessible for tending and the insulated *Energy Saver Shades* can be raised and lowered. In addition, the pattern allows varying views through the tubes to the sunny yard beyond. The glow of diffuse light into the interior is enriched by soft reflections of nearby quarry-tile floors and post and beam framing. It is a three-dimensional sculptured light box within the larger cabinetry of the "Solar Woodbox."

While sensually attractive, the water wall is also heavy and requires a substantial foundation. When full, each 8-foot tube weighs 404 pounds and holds 47 gallons of water. The foundation slab for the home is 4 inches thick everywhere except beneath the water wall where it is 8 inches thick, for both structural and thermal reasons. The slab on the lower level is paved with ⅜-inch quarry tile, which adds to the thermal mass. Beneath the slab, there are 4 inches of extruded polystyrene "blueboard" insulation. In addition, there are 2 inches of rigid insulation down the inside of frost walls and 1 inch down the outside.

The water is treated with chlorine to eliminate algae growth, and the wall's efficiency is increased by a washed-stone reflector area outdoors which is two times the height of the glass.

The glass area of the water wall is 413 square feet. All moving glass is double glazed in wood frames. Fixed panels are ⅝-inch insulating glass or sealed, tempered, double-glass units. The glass is protected by a 3-foot overhang on the upper level and movable shades on the lower level.

A Modern Post and Beam Frame

The post and beam framing of the "Solar Woodbox" is a modern version of an ancient technique. Historically, the use of heavy timbers for the framing of walls, floors and roofs is the traditional method of wood construction. European medieval half-timbered houses and pre-twentieth century American barns were always framed using post and beam construction. It is a sturdy method requiring heavy, sound timbers of thick cross section. It is also a technique that traditionally requires a practiced team of skilled jointers to cut and fit these large timbers with precision.

In the United States, post and beam construction has been replaced largely by stud-wall construction or balloon framing. This American invention of light-wood framing uses small cross-sectioned lumber such as 2×4s. It is easy to assemble and does not require the precision nor the team of human muscles of post and beam. Many Americans think that balloon framing, which goes up as fast as a balloon, is the only way to build a house.

Within the last generation, however, there has been a revival of post and beam construction. It has been rediscovered as a more sturdy method of building: it is an almost permanent, fire-resistant way to build. The regularity and the integrity of post and beam construction is appealing both to builders and homeowners who respect the importance of craftsmanship and timelessness. As the Kirleys demonstrated, it also has a practical value

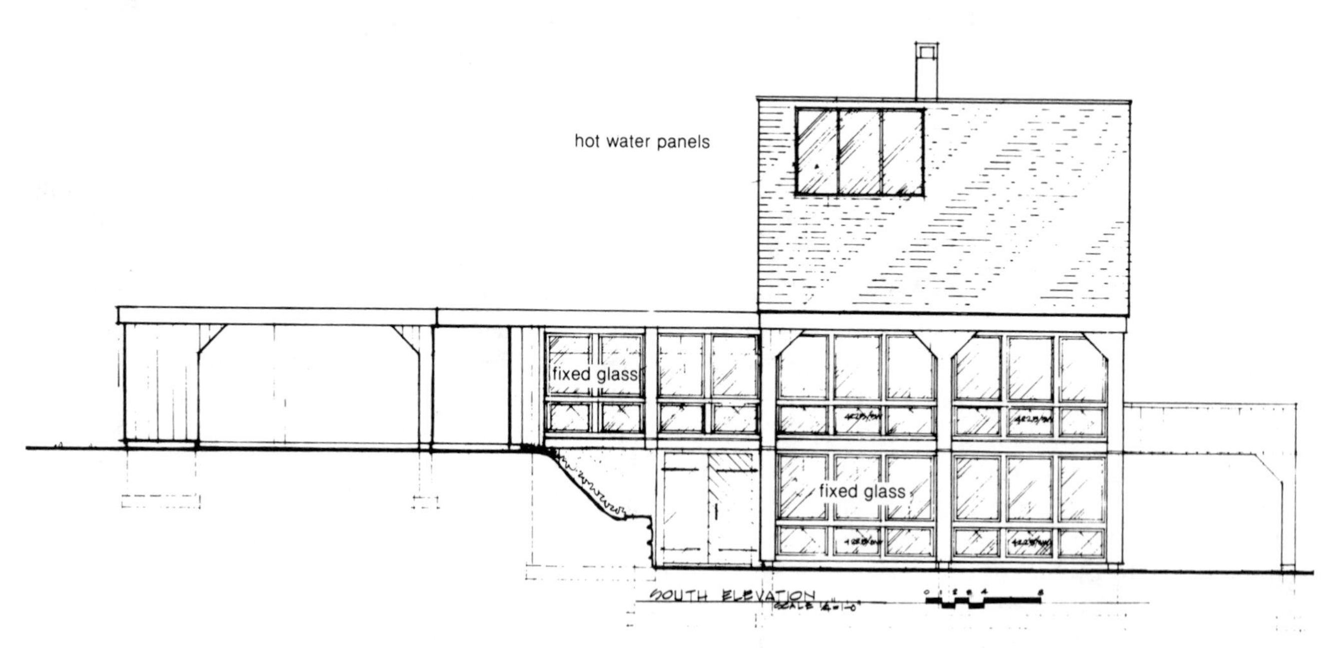

An array of windows and the solar hot water panels are features of the broad south elevation.

in passive solar construction: the beams can support the heavy thermal mass required to provide stable, comfortable temperatures.

Frame Construction

The frame is precut, native eastern white pine. Posts are solid 8 × 8s; plates and sill, 7 × 8s; and rafters, 4 × 6s. The frame was erected on site in about three days by four men and no heavy equipment. The post and beam frame not only carries the great weight of the storage wall, but allows the open, uninterrupted space necessary for radiant heat comfort and good air circulation.

Between the posts, there is a 2 × 6 stud infill panel with R-19 fiberglass batts and a continuous 6-mil polyethylene vapor barrier behind the ½-inch gypsum wallboard. Outside this system are continuous 2 × 3 nailers that act as spacers. Then the house is totally wrapped with 1 inch of *High R* insulation, cutting infiltration to a minimum. The wall has an R value of 36.

The roof rafters are 2 × 14s, allowing the use of 12-inch batts of fiberglass. This, in combination with 1 inch of *High R* on the inside, produces an R of 46.

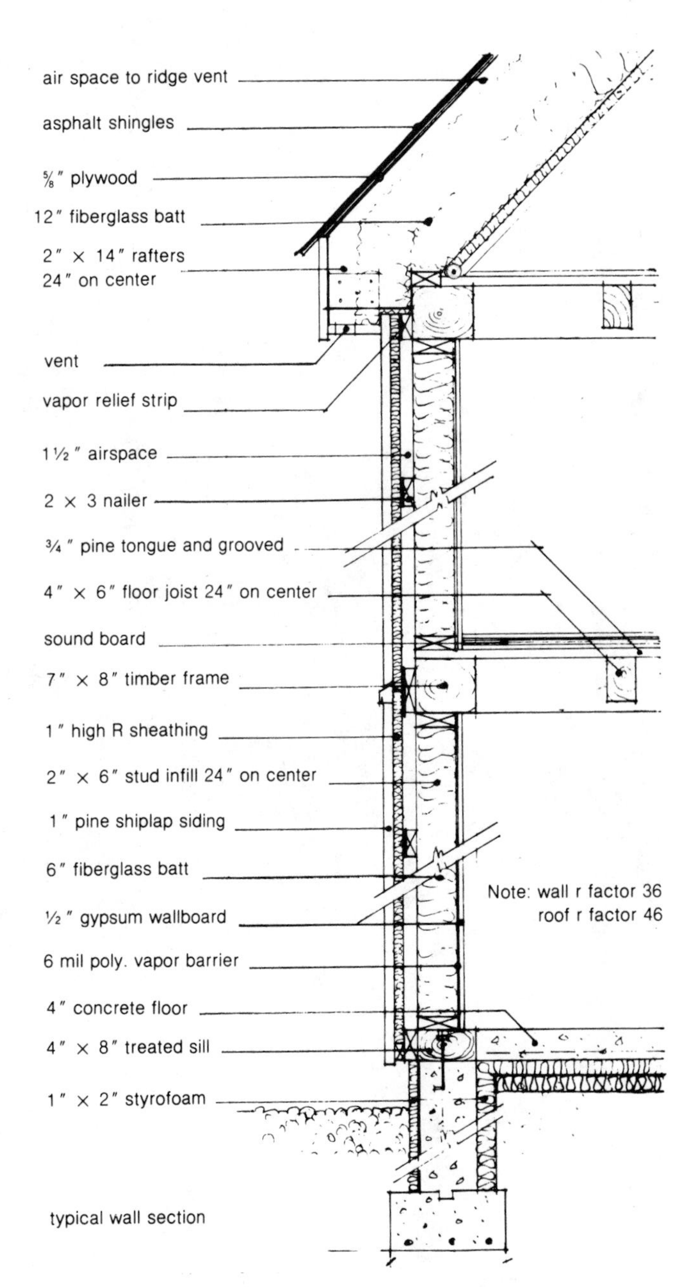

The post and beam framing permits use of heavy insulation.

The New "Solar Woodbox"

In some ways, the "Solar Woodbox" is a lean and undecorated statement of practicality. Its tight, sturdy shape is not surprising for New England, a region with harsh climate and a place where traditional values are strong. It expresses the ethic of permanence and the pride in craftsmanship of the owner/builder. Outside, the triangular frame stiffeners or elbow braces reinforce a sense of stability; on the inside, the muscular timber frame provides a handsome, larger frame of reference to the subservient spaces and furnishings.

Nothing is superfluous in this home; everything is arranged in its place. This is as true about the architectural enclosure as it is about the details of furniture and lifestyle. But, perhaps it is poetic that the one major innovation, the water wall that forms the core of this passive-solar heating system, has also become the loveliest and most decorative element of the house. Its glowing light and invisible radiant heat extend the integrity of the use of resources represented in this new interpretation of traditional construction techniques.

A New England "Treehouse"

RESIDENCE FOR

Mr. and Mrs. Jeffrey Klugman
Guilford, Connecticut

DESIGNER

Jefferson B. Riley, A.I.A.
Moore, Grover, Harper, PC.
Essex, Connecticut 06426

BUILDER

Richard Riggio & Sons
Melody Lane
Ivoryton, Connecticut 06442

PROJECT CO-MANAGERS

Dean Kuth

Leonard J. Wyeth
More, Grover, Harper, PC.
Essex, Connecticut 06426

STATUS

Occupied, Summer, 1980

Estimated Building Energy Performance

BUILDING DATA

Heating degree days	5,897 DD/yr.
Foundation area	725 sq. ft.
Total conditioned floor area	2,310 sq. ft.
Building cost	$98,000
Cost of special energy features	$10,800
Total building UA	936 Btu/°F. hr.
Heating	
Area of solar-heat collection	752 sq. ft.
Solar heat	43 percent
Auxiliary heat (oil fired furnace)	23 percent
Auxiliary heat (wood stove)	30 percent
Auxiliary heat needed	285 gallons oil plus 3½ cords of wood
Cost of auxiliary heat	7¢/sq. ft./yr. (oil $.56/gallon, wood free)
Cooling	
Auxiliary	fan-assisted night flushing
Cost of auxiliary cooling	1¢/sq. ft./yr.
Hot water	
Need	50,000 gallons/yr. @ 120°F. (incoming 50°F.)

ANNUAL ENERGY SUMMARY

Nature's contributions	
Heating	46.7×10^6 Btu
Cooling	no estimate
Hot water	13.1×10^6 Btu
Auxiliary purchased	
Electricity	79.2×10^6 Btu or 34,300 Btu/sq. ft.

This house rises like an exotic tree in the lush, New England woods. The four-story structure has a radically different shape, dictated by its extraordinary, passive solar design. In effect, the whole house is a solar collector, featuring a steeply pitched south roof that is about 80 percent glass. Inside, spaces connect by means of balconies and windows to create a dramatic, bright, expansive interior.

The Klugman residence sits below a steep hillside to the east and has a sweeping driveway that approaches through the forest from the west. There is no garage or carport. A parking court is the only formal organization in an otherwise natural landscape. Cars are dwarfed by both the house and surrounding trees. The wood exterior as well as the form of this special house fit nicely in the towering woods. In a sense, this is an inventive, year-round treehouse.

The house rises squarely from the ground without transitions. On the south side, a ceremonial stair rises in a straight run up to the outdoor deck in front of the second-story living room. Otherwise, the house is a sculptured, vertical solid. There are no visual frills to disturb its modeled surfaces. For instance, stairs and deck do not have open balustrades, but solid skirts—to add to the sense of density. Only the nearby thin tree trunks and their leaves have the softness of filigree.

At ground level, the only symbol of an entry is a large, curved arch that opens to a dark shelter under the house and porch. Inside the shelter, in the shadows to the left, is the main entry. Two bedrooms, a darkroom/bathroom and a furnace room complete the ground floor. A stair with two sets of winders rises from a hall up along the north wall to the open continuities of the main floor.

On the main floor, magnificent views of the towering nearby trees are possible. The kitchen, while open to both the dining room and greenhouse, has a completely enclosed pantry for food storage. Both kitchen and dining room acquire most of their light and character from the greenhouse to the south, although each has side windows for cross ventilation and light. The living room, with its soaring ceiling, is the surprise space of the house; a stove serves as an organizational focal point. Large windows face northeast toward a stand of hemlocks; a row of sloped, solar windows in the roof, two stories overhead, faces south. All lines and planes lead upward. The roof windows and northwest windows converge at a small, triangular ceiling, which is pierced by the vertical, metal stove flue. Adjacent balconies and 19th century stained glass windows help link the living room to other parts of the house and to memories of other places.

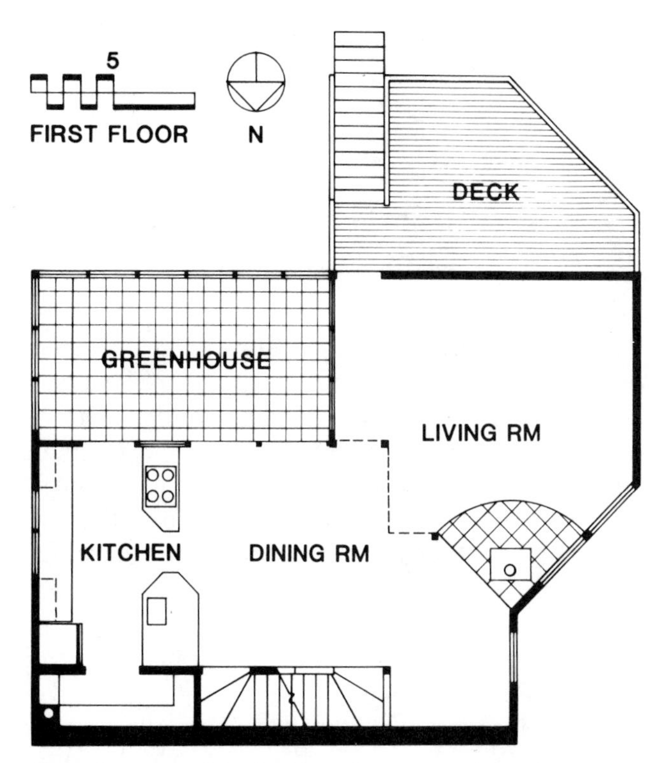

Greenhouse and deck take advantage of southern exposure.

The upper floors, with balconies and interior windows, share the light and heat of the adjacent greenhouse and living room hall. The top level is perhaps the most exciting; it is a crow's nest, especially useful as a master suite. It is accessible from a small, private stair from the large bedroom below. And it contains both a bathroom loft with balcony overlooking the living room and sliding doors out to a southern balcony four stories up. Toilet and shower are discreetly enclosed. A second connecting loft, also on the top level, has its own balcony open to the greenhouse and its own private stair up from the bedroom below. The interior has the effect of an assemblage of small houses with the qualities of a village.

An Integrated Heating and Cooling System

"The entire north wall (four stories high) is made of concrete blocks and insulated on the exterior with 3 inches of sprayed-on foam," reports Jefferson B. Riley, the architect. "In the winter, solar heat enters through the south-facing glass roof. The heated air rises by convection to a tightly-sealed and well-insulated attic plenum. The solar-heated air is then drawn by the furnace fan down from the attic plenum through the cores of the concrete-block wall, which gradually heats up during a sunny day and reradiates the heat back into the house at night. Heated air from the living room also rises up to the attic plenum to be circulated by the furnace fan. When needed, the oil burner of the furnace kicks in. In the summer, the process is reversed. A large exhaust fan in the attic pulls cool air up from the basement through the concrete block wall at night, thus cooling down the wall in preparation for it to absorb the heat of the next day. There is no refrigerated air conditioning. Three active solar collectors mounted on the roof provide hot water."

Although the north wall is critical to the effectiveness of the solar system, it is an unobtrusive component. Not only is it indistinguishable visibly

The winter sun penetrates deeply into the house, warming air which rises to the attic plenum. There the air is pulled down through the core of the masonry wall as return air to the furnace which recycles it throughout the house. As night approaches the masonry wall has absorbed much of the warmth from the solar-heated air.

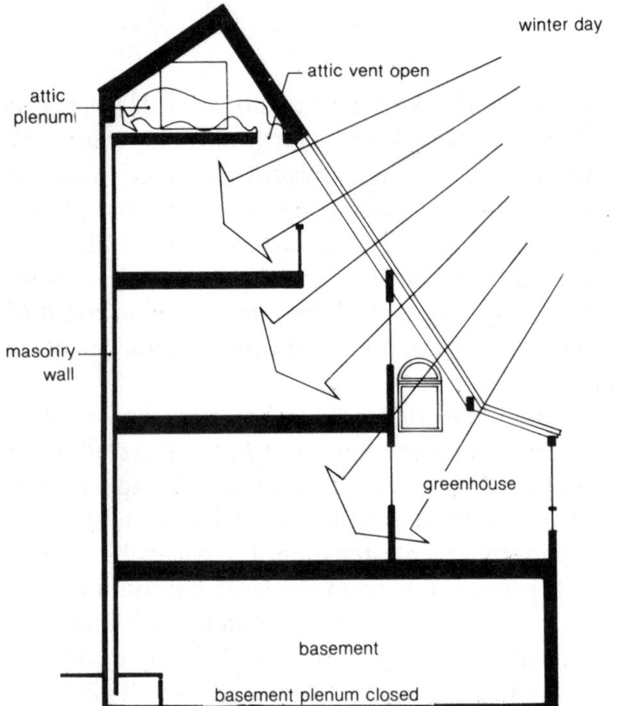

At night the masonry wall, now warm from the passage of the day's solar-heated air, radiates its stored heat into the rooms of the house. Supplementary heat is supplied by the furnace.

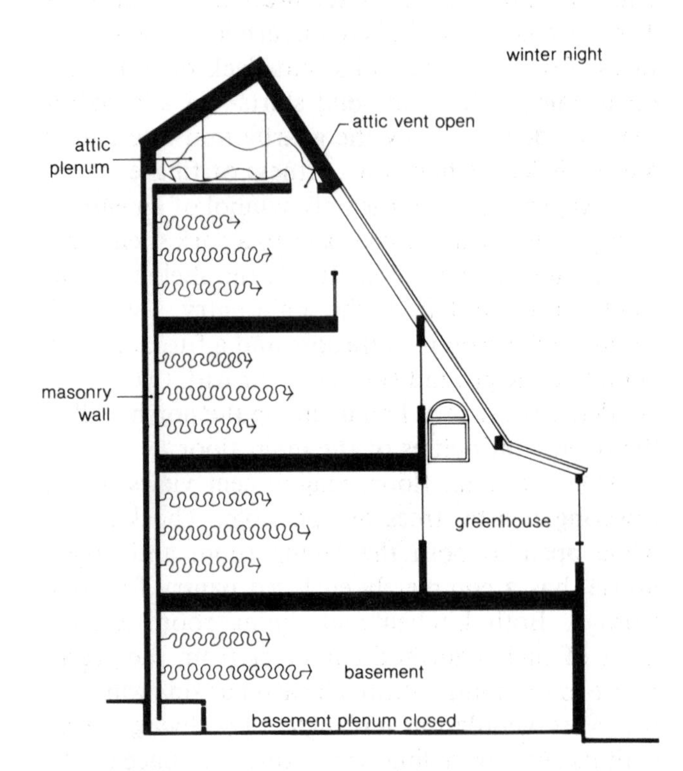

During the day, left, heat enters through the glass roof, rises to the plenum, and is pulled down through the concrete-block wall. Heat is radiated back from the wall at night (right).

There's a hint of the best of two centuries of house design in this treehouse.

from other parts of the interior, it can hardly be perceived thermally. Because it operates at relatively small temperature differences and because it has a slow temperature response, its thermal effect on the interior is difficult to sense. But, these small thermal differences influence interior comfort significantly.

In sum, this house is a personal imaginative response to a majestic environment. At the same time, the house combines character and practicality. For example, the charming appearance of the greenhouse belies its important solar contributions. With mullioned windows and old wood chairs, the greenhouse has a nineteenth-century character. A certain nineteenth-century sensitivity to precision craftsmanship and exposed wood surfaces carries this house in a different direction from the vinyl and latex surfaces of typical modern houses. But its personality even without people or furnishings stops short of the whimsical. In fact, the double hung, double-glazed windows are standard, factory-sized units. Similarly, exterior siding is a stock flushboard and the stud-framed walls were built with stock-in-trade methods. If there is any sense of the irregular, it is achieved with the most regular of means. Here, imagination has stretched the concept of a "treehouse" to include climatic and solar criteria in an inseparable, new kind of whole.

Detroit Edison Demonstration House

PROJECT FOR

Detroit Edison Utility Company
Troy, Michigan

DESIGNER

Edward Mazria, A.I.A.
Marc Schiff, A.I.A.
Thomas Cain, Job Captain
Greg Presley, Construction Supervision
Mazria/Schiff & Associates, Inc.
Box 4883
Albuquerque, New Mexico 87196

BUILDER

Bing Construction
2930 South Telegraph St.
Bloomfield Hills, Michigan 48013

STATUS

Original unbuilt design, 1980
Modified design completed, 1981
Open for viewing for one year
Sold in 1982

Estimated Annual Building Energy Performance
Preliminary Design

BUILDING DATA

Heating degree days	6,228 DD/yr.
Floor area	2,120 sq. ft.
Building cost	$100,000
Cost of special energy features	$15,000
Total building UA (day)	643 Btu/°F. hr.
Total building UA(night)	514 Btu/°F. hr.
Heating	
Area of solar heat collection glazing	530 sq. ft.
Solar heat	60 percent
Auxiliary	heat pump (COP 3.0)
Auxiliary heat needed	6,033 Btu/sq. ft.
Cooling	
Auxiliary cooling	heat pump (COP 3.0)
Auxiliary cooling needed	2,315 Btu/sq. ft.
Hot water	
Need	30,000 gallons/yr. @ 130°F. (incoming 50°F.)
Solar collector area	90 sq. ft.
Solar fraction	70 percent
Auxiliary cost	$90/yr.

ANNUAL ENERGY SUMMARY

Nature's contributions	
Heating	54.9×10^6 Btu
Cooling	not calculated
Hot water	14.4×10^6 Btu
Auxiliary purchased	
Heating	12.8×10^6 Btu
Cooling	4.9×10^6 Btu
Hot water	6.2×10^6 Btu
Total electricity (including lights and appliances)	29.8×10^6 Btu or 14,083 Btu/sq. ft.

As fuel costs rose rapidly in the 1970s, designers and homeowners were not the only ones who looked toward solar applications to cut energy costs. Electric utilities began to sponsor solar demonstration projects to increase public awareness of the benefits of solar energy applications. The solar demonstration house initiated by Detroit Edison is one of the most handsome utility-sponsored houses.

Not surprisingly, the Detroit Edison house uses electricity for auxiliary heating. But, the main purposes of the house were: to demonstrate several passive solar heating and thermal storage techniques; to incorporate passive cooling methods; and to use an active solar system for domestic hot water. To fulfill all these purposes, two demonstration projects were developed: an unbuilt, 2,120-square-foot, prize-winning design and an expanded 2,800-square-foot home, which was built, displayed and sold.

Specific Design Goals

It was important in the design that the Detroit Edison house be comparable in cost and standard with adjacent development so that it be a suitable prototype for builders and developers. For this reason, the house was designed so it could be placed on any typical 75-foot lot with four different orientations.

Because the builder was selected before the architect, he provided critical input to the design program. The builder's models and prices, just below $100,000, were a guide in creating a custom design for this market. The house was to be wood frame, to be spacious but not too large in size, and to be two stories. The two-story configuration provided greater thermal economy because of its compact form, and conformed to the market preference. A brick veneer, two-story house with pitched roof is traditional in the midwest.

One architectural goal was the optional use of various roof forms, which would allow the basic house to take on different appearances and be adaptable to different designs. As is typical of their design process, architects Ed Mazria and Marc

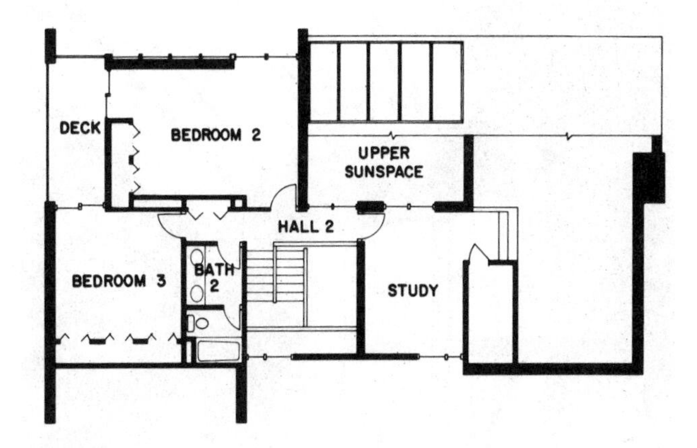

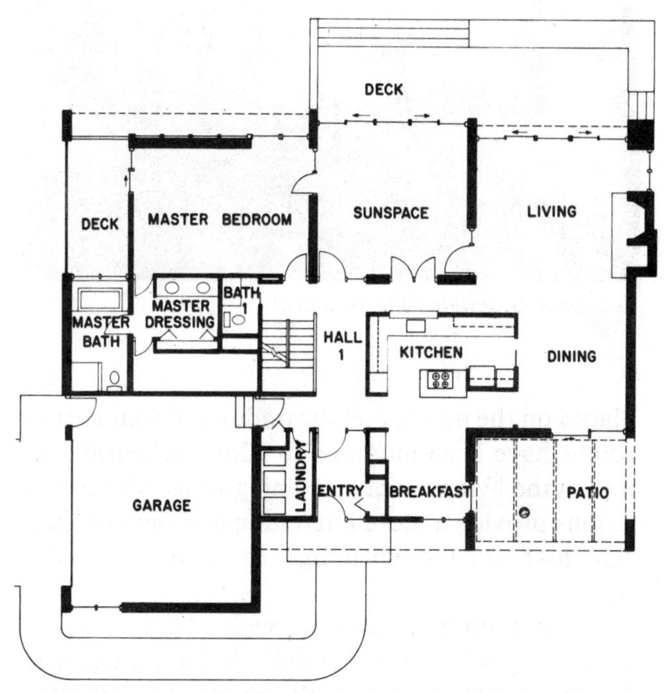

The two-story sunspace is a spectacular feature of both the second (top) and first floor plan.

Schiff used cardboard study models to develop the basic designs and to refine ideas.

Elements of the Finished Design

The design is a variation of the traditional two-story house. But, the central stairway has been

Sunspace is delightfully practical.

placed on the north, and the traditional four corner rooms have been modified to allow cathedral ceilings in the living room and dining areas. These variations provide a more interesting and flexible interior, and add the advantages of good air circulation.

The main entrance is through a large entry-air lock on the north side that links the garage with the house. In adaptations of the design, the entrance could be through the greenhouse on the south side.

The kitchen, typically a heat producer, is in the middle of the north side. Living and dining areas are separated by the masonry mass of the large fireplace. Both are two-story volumes that provide soaring interior space and help to move warm and cool air throughout the house.

A Spectacular Greenhouse

The most spectacular part of the house is the two-story greenhouse with its sloped glass roof. Like a solar furnace, it is the major source of interior heat. The greenhouse has three solid walls of 12-inch brick with concrete-filled cores. This thickness modifies the potentially high temperatures in the greenhouse and the thermal mass acts as a constant and stable heat source for the adjacent spaces.

Because the greenhouse is located in the center of the south side, it allows maximum radiant heat transfer into the main rooms. Aesthetically and functionally it connects the interior spaces and brings light into the interior.

Greenhouse heat is also stored in a 500-cubic-foot rock bed under the dining room floor, which is actively charged by hot air drawn from the top of the greenhouse. The rock bed warms the dining room floor on the northwest corner of the house, indirectly heating the interior.

Several Passive Heating Approaches

Other areas of the house have passive solar collection and heat storage approaches of their own, each selected as it related to the function of the space. The heating needs of the living room are met by using the direct gain from the south-facing windows that look out on the neighborhood. This solar gain is partially absorbed by interior brick walls and by the masonry floor, which is a 4-inch concrete slab surfaced with 1-inch quarry tile.

Both the master bedroom downstairs and bedroom number two upstairs have direct-gain south windows in addition to Trombe walls or solar thermal storage walls which help to minimize nighttime temperature drops. Using these walls as thermal storage also allows the bedroom floors to be carpeted.

Cooling the House

During the summer months, the thermal mass and the manually operated glazing act as the passive cooling system. Cooled by night ventilation, the thermal mass provides cooling for the interior during the day. Night ventilation occurs as warm air is exchanged for cool night air through low operable windows and high vents in the greenhouse and stairwell. Operable shades on the high sloping glass keep the sun off the interior mass walls of the greenhouse. And on the north, east and west, deciduous trees have been planted to further minimize unwanted solar gain.

Energy Conservation Built In

Careful design has also helped to keep heat loss in the Detroit Edison house to a minimum. The 530 square feet of solar-collecting south glass are double glazed, and windows on the east, north and west are triple glazed. An interior insulating curtain on the large living room window has a resistance value of R-10; the curtain is automatically opened in the morning and closed at night by a light-sensing device. Additional movable insulation on other south-facing windows could reduce their nighttime heat loss.

The exterior walls have 2 × 6 stud framing, allowing 6 inches of fiberglass batt insulation and plasterboard interior that give the wall a resistance value of R-20. The roof exterior has conventional ¾-inch plywood decking and asphalt shingles; below the decking, 15-inch fiberglass batts are rated at R-50. The perimeter of the concrete foundation wall is surrounded by 2 inches of rigid insulation. The result of these conservation strategies is a house with a daily heat gain/loss factor of 7.8 Btu for each square foot of floor area per degree day.

House Built—With Modifications—in 1980

The Detroit Edison Demonstration House design was accepted by the utility and the builder as a successful response to the program goals in the fall of 1980. It was this buildable design that was recognized with a National Passive Solar Design Award. However, a series of changes in the economy and in the Edison demonstration program led to several modifications in the design that was built.

Because a different house lot was selected in a more expensive subdivision, the house was upgraded in size, materials, appearance and price so it would conform with the standards of the surrounding neighborhood. Thus, the wood siding of the exterior was changed to brick veneer, a significant modification in cost and appearance over the original, although this change did not significantly add to the heating or cooling performance. In addition, the chimneys, which in the original design were light, insulated metal tubes, were built with full brick work, adding weight and dignity to the appearance of the exterior.

Floor space was increased from 2,200 to 2,800 square feet. A third bedroom was included upstairs. A completely finished basement was added, and the garage was made part of the house rather than a separate structure. The kitchen was moved to a central location on the first floor near the dining area, and the fireplace was turned to the outside wall between the living and dining areas. In spite of the many architectural changes, the original thermal specifications and design were retained, an interesting revelation of the versatility of the solar design concept.

With inflation, the cost of the house almost doubled. Part of the extra cost was an impressive level of craftsmanship in the interior finishes. Precise carpentry, neat masonry and refined cabinetry added a quality rarely seen even in custom-built houses.

In appearance, the Detroit Edison Demonstration House, as built, is certainly more substantial than the competition design. A heavy chimney and a relocated garage emphasize the massive substance of the house and add to its credibility as a desirable image. The house was admired by more than 150,000 visitors before it was sold.

It is a tribute to the architects' creativity, flexibility and careful planning that the design could be expanded so easily without losing its solar integrity. As a solar demonstration, this project developed from a desirable design with middle class economy to a home for those who can afford almost any residence of their choice.

A Snug Retirement Home

RESIDENCE FOR

Mr. and Mrs. John Black
Veedersburg, Indiana

DESIGNER

Clif Carey
506 West John St.
Champaign, Illinois 61820

BUILDER

R. L. Mueser Construction
RR 2
Veedersburg, Indiana 47987

STATUS

Occupied, July 1982.

Estimated Building Energy Performance

BUILDING DATA

Heating degree days	5,600 DD/yr.
Floor area	1,520 sq. ft.
Building cost	$85,400
Cost of special energy features	$5,460
Total building UA (day)	228 Btu/°F. hr.
Total building UA (night)	142 Btu/°F. hr.
Heating	
Area of solar-heat collection	456 sq. ft.
Cost of auxiliary heat	unknown
Cooling	
Cool tubes (three 6-inch diameter pipes, each 100 feet long)	
Auxiliary cooling	none provided
Electricity required for fan venting	$195/yr.
Hot water	
Need	16,950 gallons @ 140°F. (incoming 54°F.)
Provision for future solar system	
Present electrical cost	$203/yr.

ANNUAL ENERGY SUMMARY

Nature's contributions	
Heating	19.8×10^6 Btu
Cooling	not calculated
Hot water	none at present
Auxiliary purchased	
Electricity	46.7×10^6 Btu or 34,645 Btu/sq. ft.

This two-bedroom retirement home on 100 acres of rolling Indiana farmland benefits by nestling into a south-facing hillside. The earth shelters the home from northern wintery blasts and keeps it comfortable in muggy, summer heat. The earth's cooling effect is augmented by extensive underground air pipes buried near the house. These cool and dehumidify ventilation air for summer comfort.

These earth sheltering effects are combined with a rhythmic undulating solar south wall. Since rural views of fields and trees down the creek valley are to the south of the house, sections of passively-heated masonry wall alternate with direct-gain windows. Thermally, such a two-fold solar wall provides instant heat from the sun shining directly through the windows and delayed radiant heat from the masonry (Trombe) walls. While the Trombe wall also collects solar heat instantly, the delivery is later because of the time lag of conduction through the mass of the thermal storage wall. Thus, the Trombe wall delivers most of its heat through the winter evening, after the sun has gone down and before bedtime.

The Trombe wall sections face 7° west of due south, an orientation calculated to optimize performance for this particular location, Veedersburg, Indiana. The 6- × 8-foot windows are sliding glass doors, oriented 38° east of south, to catch the full impact of the direct sun early in the day. This allows the house to "wake up" earlier than if all the windows faced due south.

The south southeast orientation of the solar wall is also ideal for views into the nearby creek and valley. The alternating plan directions have the effect of lengthening the time of day of solar access and thus of increasing solar effectiveness. By integrating scenic and comfort considerations, the architect has designed an interesting and handsome double-angle, double-function wall.

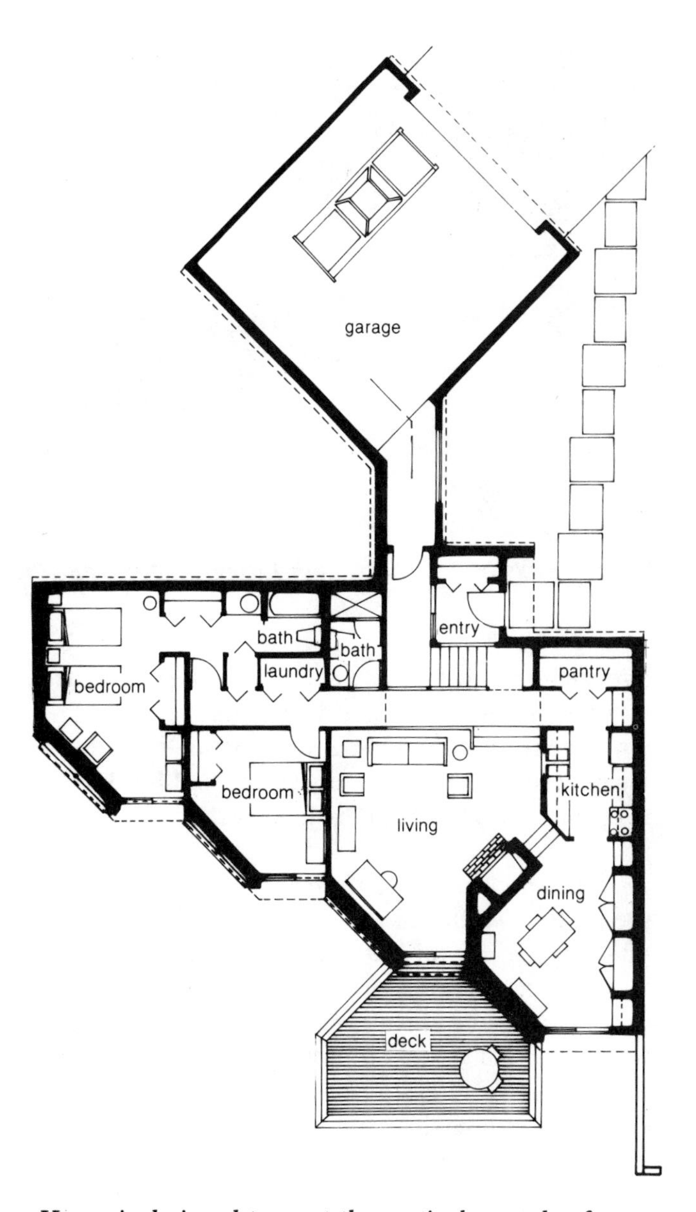

House is designed to meet the particular needs of a retired couple.

Trombe Walls and Insulation Curtains

In western Indiana, seasonal weather patterns vary considerably, in both summer and winter, so extra design efforts were necessary to insure continuous comfort in the home. Special materials and details were used to achieve maximum effectiveness from the Trombe walls, including the installation of insulated curtains.

The Trombe walls are constructed of 8-inch concrete block grouted solid plus an inner 4 inches of face brick. By providing lintels on the tops of the Trombe walls, they become non-bearing and thus could be considered for energy tax credits. The solar collection faces have a patented "selec-

Trombe walls behind these south-facing windows soak up heat on sunshiny days.

tive surface" to increase solar radiation absorbed and also reduce radiation back out the wall. The material is an adhesive-backed copper foil with a black chrome film and a nickel middle film. Its absorptivity of 95 percent is higher than any black paint.

In this climate, passive solar heat collection in the winter would be quite modest without disciplined use of movable insulation shades or curtains at night and during stormy days. Thus, self-inflating, roll-up curtains are installed in the 12 inches between glazing and Trombe wall. These curtains, with a core of multi-layered reflective foil, are operated by pull chains from the interior.

The insulated curtain also controls convection air moving into the house interior from the solar-heated air space between the Trombe wall and the glass. In the daytime, warm air rises and flows into the rooms through high vent holes that are 2 square feet for each 8- × 8-foot Trombe-wall section. Return vent holes near the floor allow a convective air loop to develop. At night, when the curtain is unrolled into place, both vent holes will be closed and thus the colder air in the space next to the glass will not flow into the rooms.

During the overheated summer period when heat gain in the house should be minimized, the reflective curtains are also used—but to keep heat out rather than in. Because of the thermal control possible with such an insulated blind, roof overhangs above the Trombe wall have been kept to a minimum. This allows the house to be heated passively through a cold spring. But, solar heat during a warm autumn can be reflected by lowering the curtains.

Another subtle construction detail that allows the house to be thermally "trimmed" according to the season is the venting of the air space between the glass and the insulated curtain. A movable insulation plug at the top of the space allows heated air to be drawn into a plenum in the attic space from where it may be exhausted. A thermostatically controlled exhaust blower may be added later if necessary.

All four Trombe wall sections are glazed with double-skinned acrylic sheets, 6 feet wide by 8 feet high, set by screws in treated wood frames. These sturdy frames are hinged to allow both sides of the glazing to be cleaned as well as access to the outside of the 12-inch Trombe wall.

Bermed Walls

Part of the economy of this earth-sheltered design is that there is no heavy earth on the roof. However, walls to the north, west and east are bermed up to the eaves. Also, earth from the house excavation has been used to build up the contours on the northwest and to fill in the ground level between the garage and the northwest house wall.

The construction details of the earth-bermed walls were developed from the calculations of a soils-engineer consultant. They are reinforced 12-inch poured concrete walls. Because of the depth of the earth cover, footings could be placed directly below floor level.

The wall waterproofing consists of two coats of cold-mop asphalt emulsion with 6-mil polyethylene embedded in each coat. The 2 inches of rigid insulation act as the protection board for the full 8 feet of wall height with an additional 2 inches of insulation for the top 4 feet of wall. Washed gravel is backfilled from footings to 12 inches from finish grade. A 6-inch diameter foundation drain pipe runs around the entire structure, with five additional 6-inch drains under the floor slab. All drains have a gravity pitch to four separate outfall positions that are protected with pest screens.

The Specialness of a Retirement House

Retirement houses have several special characteristics and features which make them different: since lifestyles are mature, the particular functional needs and relationships are well developed. Zoned uses and privacy are fundamental needs. For older people like the Blacks, there should be space for a lifetime of possessions. And, since they spend more time at home, interior spaces should be large enough and various enough to provide a comfortable and stimulating environment. Since retirement implies reduced and/or fixed income, operating costs should be minimal; the house should require little maintenance; and monthly expenses that could inflate, such as electricity and fuels, should be minimized. The latter requirement is fulfilled by passive solar designs, and all of these needs are well accommodated in the Blacks' design.

Mrs. Black can pursue her hobbies, knitting and sewing, in almost any part of the house, while Mr. Black's avocations tend to be stationary. He can continue his audiology research in the dining room where there is a unique set of study carrels that can be hidden by full-length doors.

The kitchen is at once a pivotal location, but also away from other traffic patterns. The kitchen can't be viewed easily from the living room. But, sliding shutters over the sink provide a window into the living room, to allow the homeowners to see and talk with guests.

Descending Into a Sunny Space

Entering this house is a particularly pleasurable experience. You feel as though you are first descending into the earth, then moving on, through the living room to a magnificent, panoramic view. As you arrive at the north side of the home, the contours of the earth and the pattern of low slung roof lines draw your attention downward, down to the earth below and the sunken entrance. To enter, you step through either the insulated garage or the formal entry; both, by the way, act as protective air locks.

And both lead to the entrance hall. The hall also serves as an overlooking balcony, from which you can glimpse the views that lie ahead. Descending down a half level, you arrive at the main floor. Down another three steps, and you reach the sunken living room. The full height of this large space and its imposing command of the landscape are finally understood. And, after descending through this "hill inside a house," you can move forward again, onto the outside balcony for an unobstructed, pastoral view.

In another dimension, the plan itself is a kind of horizontal cascade. Within the house itself the floor is flat. But behind the undulating south wall each of the major spaces is quite different, not just in function, but in architectural definition and shape. All of the rooms step along parallel to the view of the creek and the natural world beyond, each providing a variation in accommodation, each providing a unique view of a very particular site.

A Modern Adobe In Sante Fe

RESIDENCE FOR

A Professional Couple
Sante Fe, New Mexico

DESIGNER

Robert W. Peters, A.I.A.
10 Tumbleweed Northwest
Albuquerque, New Mexico 87120

BUILDER

Communico, Inc.
Box 81-D
Route 9, Seton Village
Sante Fe, New Mexico 87501

CONSULTANT

Susan Nichols, solar engineer and designer
Communico, Inc.

STATUS

Occupied, Winter 1979

Estimated Building Energy Performance

BUILDING DATA

Heating degree days	6,016 DD/yr.
Floor area	1,795 sq. ft.
Building cost	$152,000
Cost of special energy features	$8,500
Total building UA (day)	476 Btu/°F. hr.
Total building UA (night)	350 Btu/°F. hr.
Heating	
Area of solar-heat collection glazings	448 sq. ft.
Solar heat	63 percent
Auxiliary fuel	electricity
Auxiliary heat needed	12,813 Btu/sq. ft.
Cost of auxiliary heat	23¢/sq. ft./yr.
Cooling	
None necessary aside from openable windows	
Hot water	
Need	9,125 gallons/yr. @ 120°F. (incoming 50°F.)
Auxiliary energy/electrical cost	$94/year

ANNUAL ENERGY SUMMARY

Nature's contribution	
Heating	48×10^6 Btu
Auxiliary purchased	
Electricity	28×10^6 Btu or 15,777 Btu/sq. ft.

The adobe look is more than just a stylish appearance for this custom-designed home in northern New Mexico. The native, sun-dried clay is discreetly combined with modern construction materials to produce a low-maintenance, energy-efficient, two-bedroom residence; up to 70 percent of the annual heating requirement is met with passive-solar heat. That's a significant achievement in Sante Fe, a 7,000-foot-high city at the southern end of the Rockies. Sante Fe has a surprisingly harsh winter climate, marked by intense snow storms and an annual heating load greater than Boston's.*

The owners, designer and builder kept the climate, topography and vegetation in mind when they situated the house on a five-acre site of arid uplands. For example, the house sits 18 inches below grade and approximately 100 feet south of a ridge line, which provides protection from prevailing winter winds. In the summer, southwest winds enter the house through vents, a means of natural cooling. The site is covered with native pinon and juniper, two low dusty-purple evergreens. Their groundcover helps preserve the moisture of the scarce rainfall. Banks of bushes planted on the north side add a protective belt of landscaping. As is typical in this rural region, there is no formal front garden or ceremonial driveway.

Combining Old and New

This home blends elements of traditional adobe architecture, contemporary solar design and standard building materials and methods. The result is a well-built home with a clearly southwestern flavor.

Walls are covered with adobe-colored stucco and built with 2 × 8 studs and 7½-inch fiberglass batt insulation. The walls have an R-value of 29, a great thermal improvement upon an uninsulated adobe wall. The flat roof has 2 × 10 joists and 9½-inch fiberglass batt (R-36). Windows are double-glazed Pella units with an insulation value of U .65. Wood doors at airlock entries have U of .34.

*Boston has an annual heating load of 5,791 degree days; Sante Fe's is 5,846.

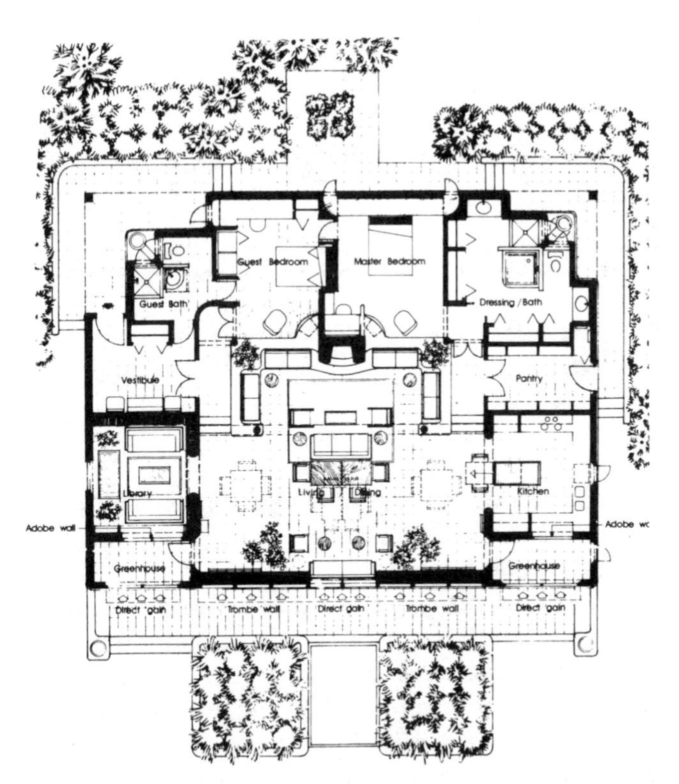

Multiple use of space is a plus of this floor plan.

Although adobe construction is especially indigenous to the area, it is an expensive material because of the cost of labor. Therefore, adobe is used here only on the interior, where its thermal storage contributes to comfortable, stable temperatures. Combined with the thick 2 × 8 stud walls, the heavy adobe interior details contribute to a feeling that this massive home is all adobe, an advantage that adds appeal and marketability to a home in this area.

A Versatile Interior

Inside, a large, central hall (36 × 25 feet) is the focal area for social occasions. Adjoining rooms can be opened to social activity or can serve other functions as needed. Thus, the kitchen may also be used for serving a buffet; breakfast may be celebrated in the greenhouse on a winter morning; the library may be used as a guestroom; the second bedroom

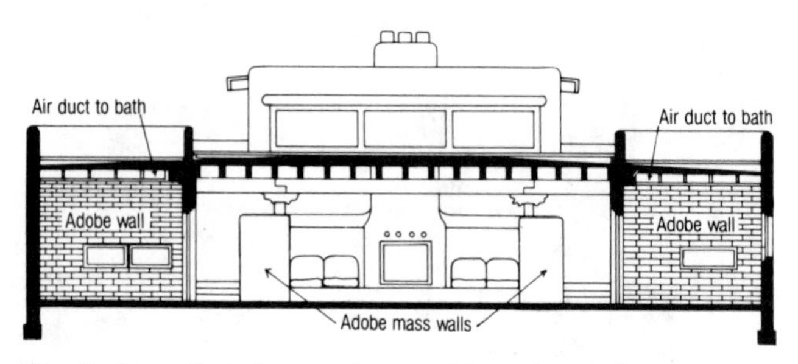

South elevation shows glass walls and broad clerestory.

doubles as a work room with a Murphy bed hinged away in the closet; and the master bedroom includes a study alcove.

The central living room is also versatile. It accommodates dining in various configurations by rearranging tables. The seating arrangement can rotate to face outward in the summer and toward the fireplace in winter. Use of banco seating accommodates a large party without making the space appear empty when the owners are alone. The banco is a broad, built-in seating platform, covered, in this case, with trim cushions.

No artificial "formal-informal" distinctions are made in this home. The occupants use all spaces according to season and mood. In the same style, the interior combines neat, modern, tailored furniture within the bold cubic spaces bounded by such primitive materials as plaster-washed adobe and rough-sawn beams. The grid of cement floor tiles ties together the clean geometries and natural colors of the house and its contents.

A Multipurpose Clerestory

High windows above the roof in a clerestory allow the sunlight to penetrate overhead into the living room and front and back zones of the house. Thus, the interior thermal mass of the fireplaces and built-in furniture is heated by the sun directly. Experience in cold climates has shown that employing interior mass as thermal storage is cheaper and more efficient than using exterior walls for thermal mass. Interior thermal mass can radiate heat in all directions, whereas thermal mass on an exterior wall is only effective on one side. Also, the problem of providing protective finish materials outside the rigid insulation used with exterior mass walls is not solved easily.

Camel-colored cement tile on all floors except in the carpeted bedrooms also adds mass and enhances the thermal stability of the interior. The thickness of exterior walls and the quantity of masonry on the interior provide an environment of stability and stillness. Because the heating is primarily passive, there are no distant fans or pumps providing background sound.

While the comfort from the interior mass is invisible, the natural lighting from the clerestory windows is dramatically visible. It provides interesting and well-balanced light in every space without sacrificing wall space. Modern paintings and prints can be displayed without the use of special artificial lighting fixtures.

A Passive Solar Collector

The house, containing 2,084 feet with 1,795 square feet of heated area, is an effective passive solar collector. Because of the well-insulated construction, the plan does not need to be elongated to increase solar gain. The home's compact, almost square form optimizes its southern exposure and passive solar gains, and minimizes exterior wall surfaces.

The south wall is a continuous series of passive solar heating components. The ends are greenhouses that provide both heat and humidity, while extending the short growing season in cold, dry, northern New Mexico. Thermostatically-controlled fans move heated air into the north bathrooms from the greenhouses. The greenhouses have interior adobe walls that collect heat; this is passed on through windows to adjoining rooms and then the remainder of the house. Light and humidity are also transferred passively from the greenhouse.

Sixteen-inch thick, concrete, thermal-storage (Trombe) walls and double glazed, direct-gain windows also provide solar heat directly to the interior. The black, unvented Trombe walls, installed behind double glass, absorb solar heat, which is conducted to the interior surfaces and then moved, by convection, to the rooms by night.

The double-glazed windows and openings on the south side have wood "eyebrows" that prevent the high (78°) summer solstice sun from reaching into the house. But, the low (32°) winter noon sun penetrates deeply, to warm the interior and add a cheerful atmosphere. In the summer, the rooms are cool and shady. Cool summer nights allow natural ventilation to remove any heat generated in the house during the day. Openable vents and high clerestories enable easy air circulation.

Although moveable insulation was part of the original design and included in the calculated estimates for winter heating, it was never installed. It has not been considered necessary. Although the house has not been scientifically monitored, the owners report a much better winter performance than anticipated. During winter, almost no auxiliary energy is used.

Lighting adds drama to interior areas.

Fireplaces and Electricity

Beyond passive solar heating, there are two back-up energy sources. There is a "heatilator" in the large, living-room fireplace and a small, beehive-shaped fireplace in each bedroom. All three fireplaces use outside air ducted directly to the firebox for combustion. Automatic back-up heating is provided by electric-resistance, radiant cable in the tile floor and an electric baseboard heater in the carpeted master bedroom and bath.

Artistic Application of Adobe

Adobe may not seem like an appropriate choice for a modern and efficient house in a cold climate. However, the traditional appearance conceals careful planning and discrete construction. The architect, Robert Peters, has developed a compact, centralized design with fine logic. The house seems even larger than its generous dimensions, and every corner is useable. The cube-like shapes of the building have a simplicity that is compatible with traditional materials and folk construction. The refinements of electric back-up heating, modern kitchen and other twentieth-century conveniences are not allowed to intrude on the elegance of the local style.

The combination of insulated wood frame with interior adobe and concrete reflects a pragmatic approach which uses mass where it is thermally efficient, and insulated frame construction where heat loss is the primary concern. Each material is used where it works best. Adobe, which is fragile when exposed to the elements, is used only in the interior where it is protected. But, the exterior weather skin is cement stucco, a material that stands up indefinitely to the freeze-thaw cycles of winter weather. Consistency of style, then, is based not just on appearances but on performance. The integrity of the solar design is an inherent part of the design sense of every part of the house. But, most important, the whole is accomplished with artistry.

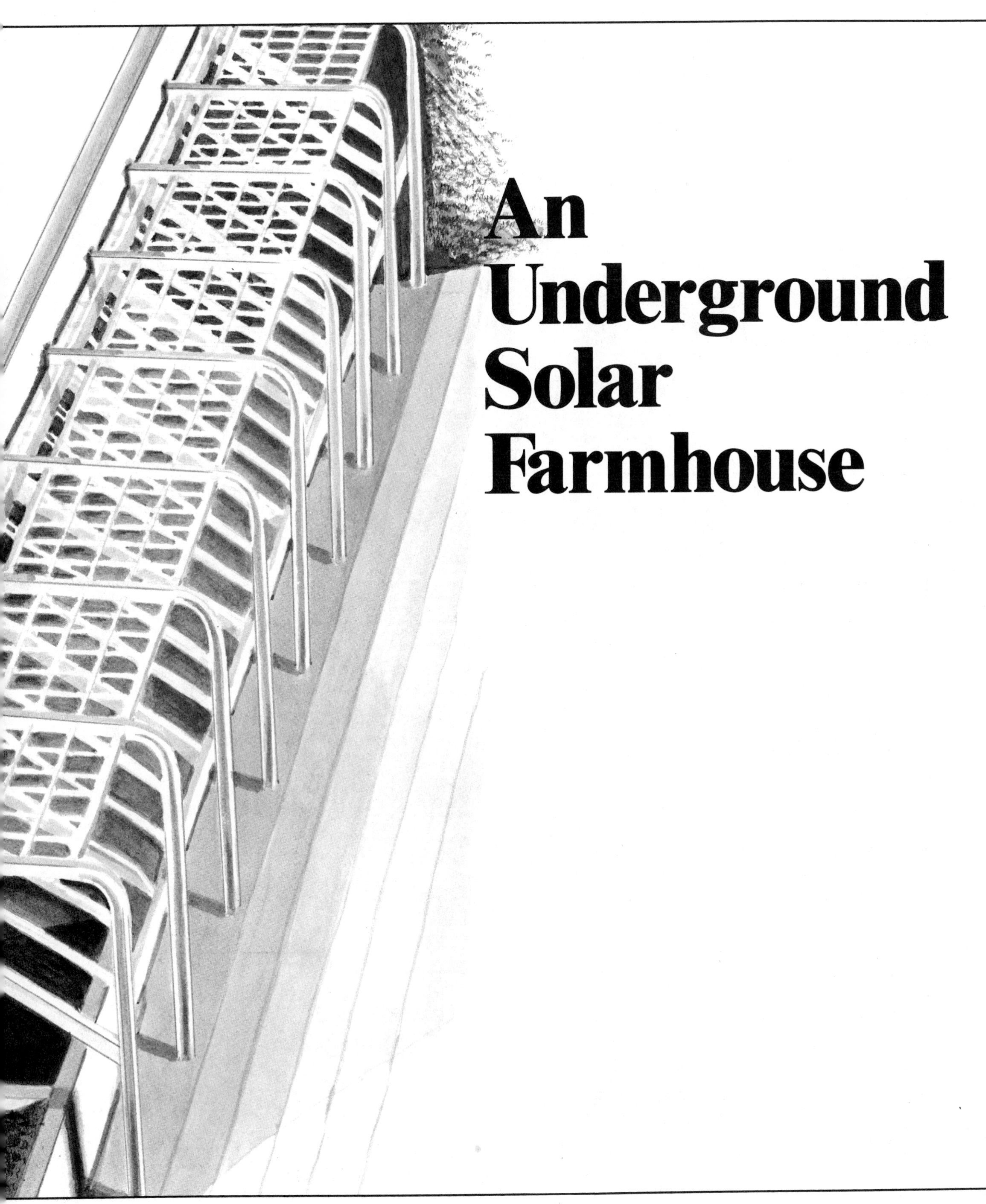

An Underground Solar Farmhouse

RESIDENCE FOR

Tim and Marilyn Hansen
Moses Lake, Washington

STATUS

Unbuilt

DESIGNER

David Miller
Miller/Hull Architects
911 Western Avenue
Room 220
Seattle, Washington 98104

Estimated Building Energy Performance

BUILDING DATA

Heating degree days	5,846 DD/yr.
Floor area	2,396 sq. ft.
Greenhouse	836 sq. ft.
Net conditioned area	1,560 sq. ft.
Building cost	$95,000
Cost of special energy features	$ 8,000
Total building UA	1,479 Btu/°F. hr.
Heating	
Area of solar-heat collection glazing	960 sq. ft.
Solar heat	87 percent
Auxiliary fuel	13 percent (wood and electricity)
Auxiliary heat needed	52,690 Btu/sq. ft./yr.
Cost of auxiliary heat	2¢/sq. ft./yr.
Cooling	
Natural means (Convection, cool tubes shading and earth tempering)	80 percent
Fans	20 percent
Auxiliary cooling needed	13,861 Btu/sq. ft./yr.
Cost of electricity for blowers	not calculated
Hot water	
Need	28,000 gallons/yr. @ 120°F. (incoming 60°F.)
Auxiliary electricity cost	$31/yr.

ANNUAL ENERGY SUMMARY

Nature's contributions	
Heating	450. × 10^6 Btu
Stack convective cooling	40 × 10^6 Btu
Hot water	8 × 10^6 Btu
Auxiliary purchased	
Electricity and fuel wood	35.9 × 10^6 Btu or 14,977 Btu/sq. ft.

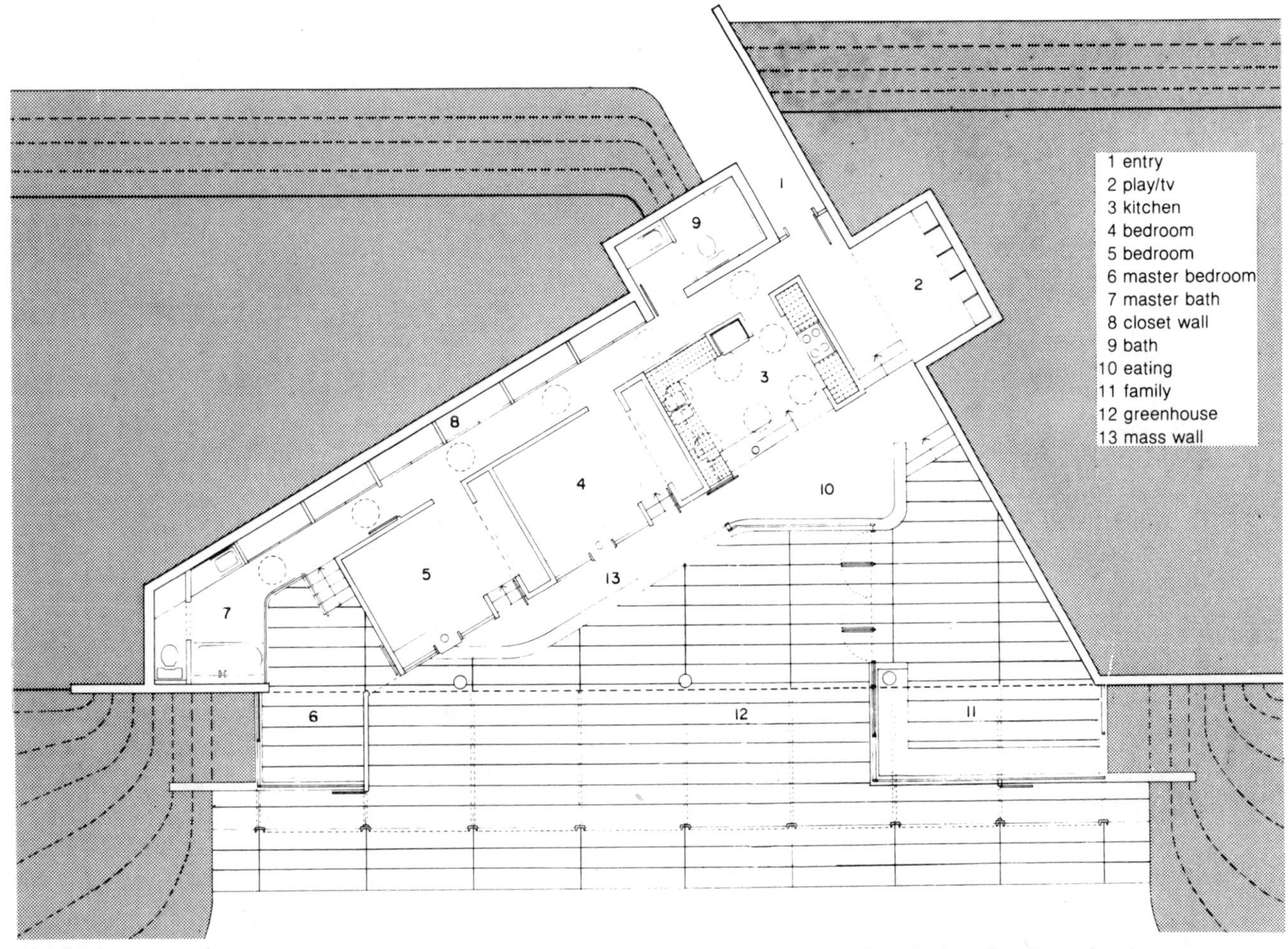

The fifty-foot long solar greenhouse provides area for social activities as well as being food-production center.

The choice of an underground design for a rural farmhouse in an extreme climate in central Washington may not need much explanation. Behind the lush coastal range, the climate of the central valleys can be harsh: hot and dry in summer and cold in winter. The potential of a greenhouse as the dominant south-facing space in an underground design may also seem obvious. But a 50-foot long solar greenhouse that could serve as the home's social focus as well as a food-production center makes this passive design special.

The 12-foot-high greenhouse, the only exposed outside skin of the farmhouse, will act as a buffer between the harsh variations of the outdoor climate and the house itself. Furthermore, the greenhouse is planned to allow wide fluctuations of conditions without adversely affecting the comfort inside the house proper.

Multi-season Greenhouse as a Passive Collector

The greenhouse is also designed to be a passive solar heat collector, with distinctive, flexible thermal-management features. These are needed in central Washington, a region with short winter days and freezing temperatures, as well as summer days with temperatures exceeding 100° F. Daily temperature swings are also large: sometimes there

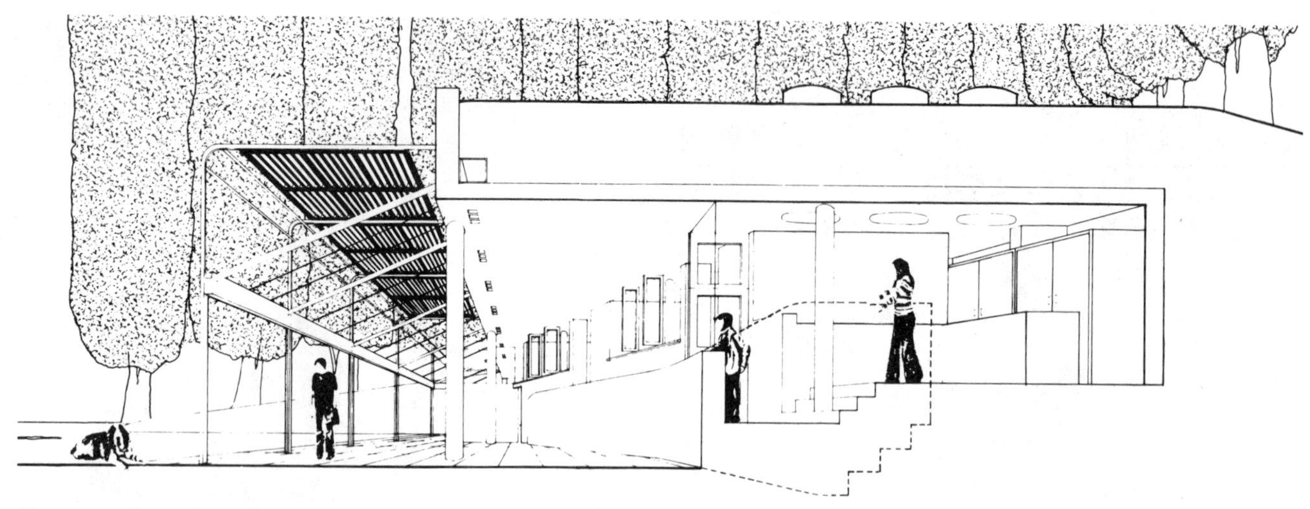

Cross-section view illustrates pleasing effect of varied levels.

is more than a 40-degree difference between the daily maximum and minimum. To meet these problems, and to prevent either winter heat loss or summer overheating, the architects selected a commonly available, movable exterior skin—wood-framed residential garage doors.

The double-glazed, factory-made doors function as a greenhouse wall in the winter to help retain heat. In the summer, they may be rolled up, opening the greenhouse to the outside. In the rolled-up position, they are part of the summer shading for the stationary glass roof. Louvers attached to the roll-up doors improve summer shading. Views from the house are unobstructed, and the greenhouse is a spacious, outside porch.

The sloped-glass roof of the greenhouse will also be double glazed, but with fixed standard patio door sealed units. The glass garage doors are moved manually since they have the balancing hardware typical of all garage doors. Typically, the roll-up doors will not need to be operated on a daily basis.

The Underground Challenge

In addition to the spacious greenhouse, this farmhouse design calls for earth sheltering, an approach that offers both rewards and challenges. Among the underground design problems, aside from waterproofing, is the obvious need for heavy structural design to hold back the weight of the earth. Especially the overhead load of several feet of wet soil can require extra structure where cost does not justify the thermal advantage. Thus, even buried buildings often combine a modest amount of earth overhead with additional roof insulation. In this case, 3 inches of rigid insulation on the roof and 2 inches of rigid insulation on the walls reduce the thermal contact with the surrounding earth. However, the floor slab would be in direct contact with the earth. Two inches of insulation on the vertical foundation wall of the greenhouse would thermally separate the earth under the greenhouse from outdoor ground temperatures.

Planning Underground Space

The earth encloses all sides of the house except the south. The house is entered from the north via a tunnel which limits wind exposure. The tunnel also expresses the architectural concern for the quality of earth protection and enclosure.

Circular skylights allow daylight to penetrate the deepest part of the interior. A large, country-sized kitchen is at the center of the plan. The two children's bedrooms have a private back hallway lined with closets and lighted naturally from round skylights overhead. Together with the bedroom doors and windows facing south into the greenhouse, the bedrooms have the two separate means of exit and egress required by fire-safety regulations. The bedrooms and kitchen have a raised view to the south, looking over the living and social spaces through the greenhouse to the views beyond. The plan has been skewed so that the enclosed rooms face an interesting open spatial continuity which is bound by the greenhouse. The changing directions and edges of the interior space help dispel the negative connotations of being "underground."

At the back of the greenhouse, the floor is terraced upward a total of 4 feet. Aside from opening views to the south from deep inside the house, this also allows the greenhouse to have a high ceiling and large south aperture. The greenhouse is terminated by the master bedroom on the west and the family room on the east.

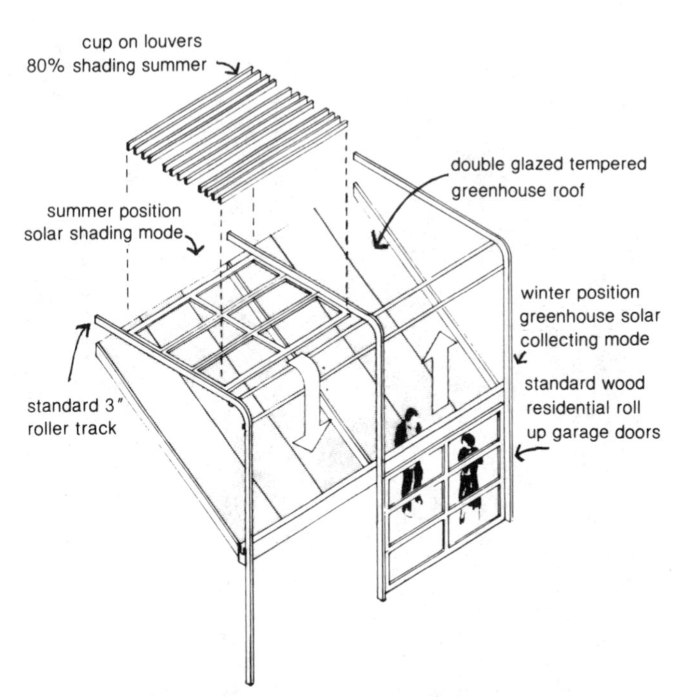

Standard residential garage door is feature of solar skin assembly.

Controlling Interior Winter Temperatures

Winter solar heat gain will be stored in two ways. Some will be stored in the concrete floors and walls of the greenhouse and in living areas exposed to direct sunlight. Because there are almost 2,400 square feet of floor slab, most of the solar heat gain will be stored in this manner. In addition, a 2-foot deep, 18- × 20-foot, rock-storage bin beneath the family room will store higher temperature air supplied by heat from the greenhouse collector duct. Another duct will return air from the bin to the greenhouse. The floor slabs are poured directly on top of the bin. Thus heat from the rock bin will provide a warm, radiant floor in the family room.

Summer Cooling

A fundamental summer cooling strategy is to avoid solar heat gain. In this house, that solution is primarily resolved within the greenhouse design. In addition, the house will be opened for flushing with cool, night air. This practice will work especially well in an arid climate such as central Washington. The 3 feet of soil over the insulated roof provide a substantial barrier to the intensity of the high summer sun. The design also includes the auxiliary use of cool tubes that draw ventilation air from outside and cool it from earth contact through an extended 50-foot labyrinth of air pipes buried 5 feet deep underground. Thus, no refrigerated air conditioning was considered necessary.

Backup Heating

A high-efficiency, 2-foot diameter *R AAS* wood stove in the family room will be the primary backup heating system. In addition, electric radiant-heating panels similar to those used to heat commercial outdoor spaces will be used for "task heating."

A Vermont Vacation Fantasy

RESIDENCE FOR

Michael Datoli
Windham, Vermont

STATUS

Occupied, 1979

DESIGNER AND BUILDER

Robert Foote Shannon,
Architect/Developer/Contractor
49 Garden Street
Boston, Massachusetts 02114

Actual Building Energy Performance

BUILDING DATA

Heating degree days	7,500 DD/yr.
Floor area (one floor)	432 sq. ft.
Total floor area	1,267 sq. ft.
Sale price (building and land)	$50,000
Cost of special energy features	$ 5,000
Total building UA (day)	342 Btu/°F. hr.
Total building UA (night)	272 Btu/°F. hr.
Heating	
Area of solar-heat collector	385 sq. ft.
Solar heat	70 percent
Cooling	
Natural means	convection, shading and earth tempering
Hot water	
Need	7,000 gallons/yr. @ 110°F.
No solar yet	
Electricity cost	$65/yr.

ANNUAL ENERGY SUMMARY

Nature's contributions	
Heating	24.0×10^6 Btu
Cooling	not calculated
Hot water	not installed
Auxiliary purchased	
Electric	½ cord firewood and 17.2×10^6 Btu or 19,111 Btu/sq. ft.

A vacation house in the mountains can either nestle into the steep slopes or soar above tricky foundations to be free of the ground. This design does both. Half buried, half flying, this vertical Vermont vacation house weathers rough New England winters on little energy. Although typical local holiday houses use almost as much energy for space heating as full-time residences, this passive solar retreat needs no auxiliary systems when not occupied. Its winter temperatures without occupants have been between 45° F. and 80° F.; a series of energy options have improved this performance substantially, by raising the lower end of the temperature range.

This is no conventional ski condominium or mountain cabin. Its flair is evident in both architectural form and in color. Its tall, narrow profile allows the best of mountain views and solar access as well as an open interior of spatial complexity. The brillant red exterior seems to relate to the red barns of rural Vermont. Its brash brightness makes a small, stimulating building seem larger and more exciting. In fact, it was inspired by a newly painted Mexican ski-lift shed. Other colors and patterns have a Mexican influence, especially the azure-blue sky coloration of the interior ceiling.

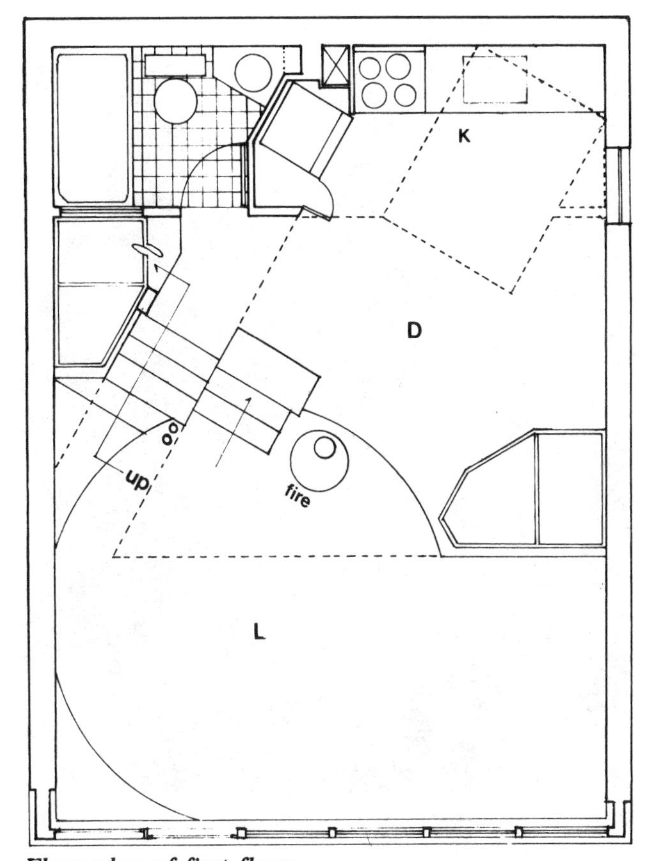

Floor plan of first floor.

An Exterior Aesthetic of Stepped Layers

The stepped layering of distinct colors and textures on the exterior is a free interpretation of classical building composition of base, column and capital. It also suggests the space organization of the interior. Next to the ground line, the wall emerges from the ground in striated concrete blocks, a kind of rusticated base. Then, smooth blocks form a transition to the hot-red wall that steps above the snow line. The colors and textures are a counterfoil to the changing colors of the native landscape.

Winging upward above the four rows of south-facing solar windows are cantilevered overhangs with shiny spaced planks that allow wind passage and keep sun off the glass in summer. But, in winter, they not only admit full sun but reflect additional solar energy into the interior. Their exaggerated eyelid profiles extend both the apparent size and the character of the house. Their 45-degree tilt provides a slight overshading for the fall season to avoid overheating.

A Speculative Solar Retreat

The design was generated and built as a speculative solar venture by Robert Foote Shannon of Boston, a seasoned solar architect and builder. He built it partially with his own hands, drawing upon both his intuitive building sense and his experience in solar calculation developed from other projects in this climate.

Its location at Windham is near the Vermont ski slopes of Stratton, Bromley and Magic Mountain. It was intended both for winter and summer

A corner of the living area, from staircase.

recreation—a weekend counterpart to the big city apartment. Located more than four hours drive from New York City, it was purchased by Michael Datoli, a New York photographer who likes the retreat so much he uses it as a subject for his work. Datoli thinks of the place as "a fantasy playground of color and space."

Passive Solar Heating with a View

The house receives direct passive solar heat only from the full-sized windows that look into the New England forest and across to the mountains beyond. Even the basement has 7-foot windows. Under the living room floor is an insulated bin with 400 cubic feet of earth for thermal storage. A pattern of manifolds provides enough surface area to change air temperatures going through the bed by 10°F. Beside it is an inside water cistern with a 1,200-gallon capacity. Rain water from the roofs is collected in the cistern and filtered, thus avoiding the need for a well. The house has separate water-supply lines, to carry recycled grey water for secondary uses.

An Interior of Juxtaposed Forms

Like the stepped layers of colors and textures on the exterior, the interior is a layered space with floors intruding into the conditioned volumes. While there are approximately four floors with levels corresponding to the rows of windows, the open interior consists of a juxtaposition of platforms, balconies, solids and cutouts that resemble a three-dimensional painting. The design was derived from Pueblo forms and colors—stacked village against textured cliffs under an azure sky. The principal level, the living room, is shaped by the circular forms of a kiva. Above the kitchen are two bedrooms, one floating above the other. One is a rectangle set at an angle, the other a sleeping platform with a free-form edge approached by a ladder.

Only the bathroom and possibly the kitchen retain some conventional pattern. Otherwise, the juxtapositions of spaces and forms are emphasized by bright colors and free contrasts of materials.

Unoccupied Building in Cold Climate

Achieving above-average thermal performance in a cold-climate building that is not occupied continuously is especially challenging. In winter, the house cycles freely, meaning it heats itself passively while the owners are away. The architect felt proud of the ability of the house to passively protect itself until a house guest mistakenly left a sliding glass door open 3 inches when departing after a January weekend. During the following eight days, the outside temperature fell to −30°F. and winds gusted up to 50 miles per hour. The temperatures in the house fell, and pipes in the lower part of the house froze. One hundred dollars of material and five hours of labor repaired the damage. It proves that designers can be as clever as they like, but buildings are always fragile against human error.

Even though the house heats itself passively, there may be too much interior thermal mass. According to the architect, when one arrives in this unoccupied house at 11 P.M. on a cold winter night, the interior does not always respond quickly to attempts to raise the temperature. He has considered having at least one bedroom that would have insulation but little thermal mass. This room then could be heated quickly using radiant electric panels.

But, a house which goes through a severe New England winter of weekend occupancy using only a half a cord of wood must be respected. When that wood is harvested right outside the door, the home's energy requirements hardly deserve criticism. In this case the owner cannot use all the wood cut from his own small lot.

Solar Details and Energy Options

The 600 cubic feet of concrete masonry exposed directly to the solar heat coming through the direct-gain windows provides stable interior temperatures. A 1,000-cfm fan constantly draws stratified hot air from the top of the house and circulates it through the rock storage bin in the basement. It operates at a cost of less than 10 cents a day.

Because of the tax advantages to homeowners who install solar hot water systems, the house was not so equipped by the builder—thus leaving the owner with the option of collecting the tax credits. But, the location of the solar hot water heater at the top of the house was planned. In addition, because the house was built on speculation, the owner was given several other energy options. At the living room level there is double glazing with R-4 night insulation already installed. But, at the two upper levels there is double glazing and room for night insulation to be added later. When the house was designed in 1977, information on the optimum sizing and characteristics of multiple glazing and movable insulation was not as conclusive as it is today.

Living area, with staircase in rear.

Vacation Houses are Fun Houses

It has often been observed that Americans can be quite conventional in the form and lifestyle of their full-time residences. But, the architectural accommodations and the patterns of lifestyle in the summer cottage, the hunting lodge and the vacation resort can be different. Often, the more stimulating their variance, the better. This lively Vermont vacation house has plural experiences in the variations of each floor—ranging from cave shelter to crow's nest. With a foundation area of less than 500 square feet, developer-architect-builder Shannon has provided the environments of many vacation worlds in which snug passive solar design is incidental to the spirit of fun.

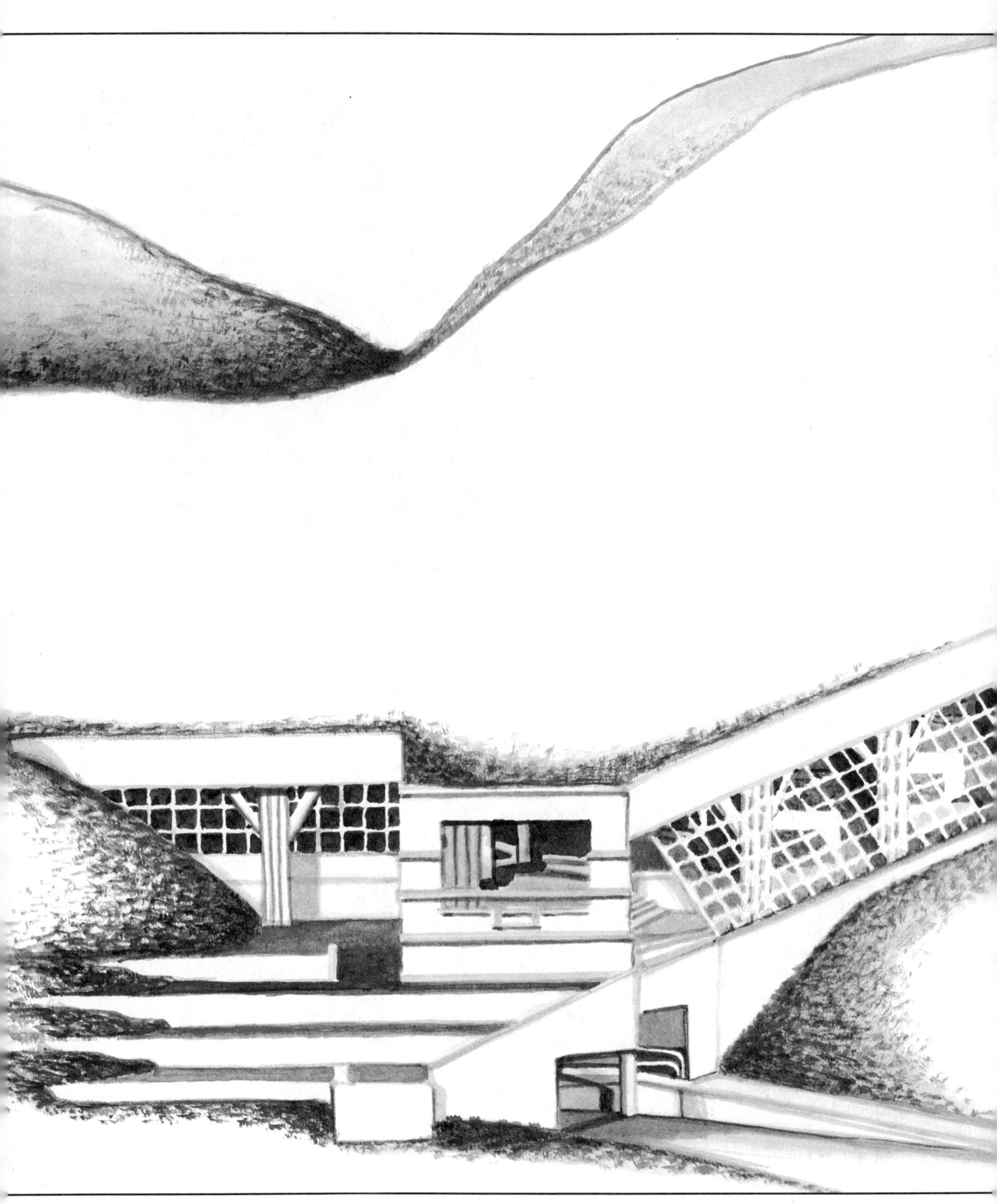

Stepping Up A Mountainside

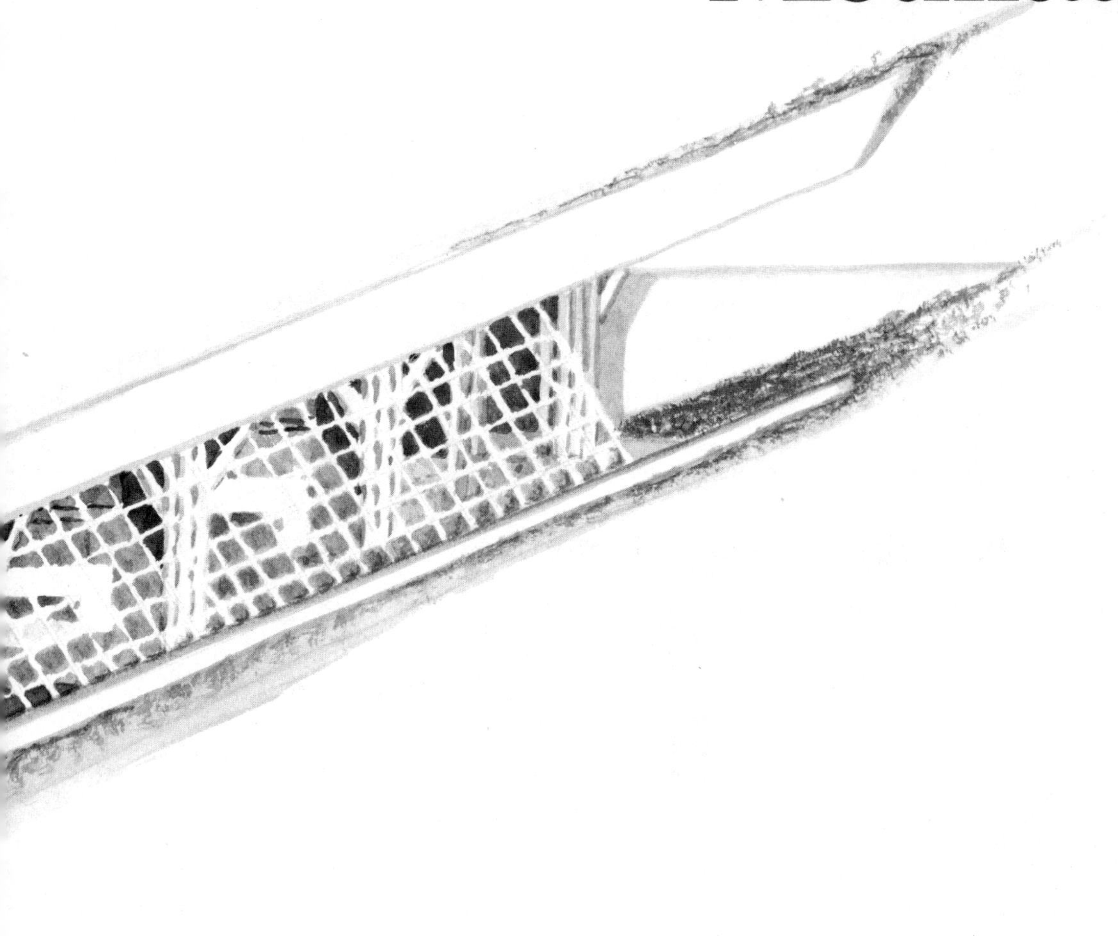

RESIDENCE FOR

Barney and Nancy Kimmick
Cle Elum, Washington

STATUS

Unbuilt

DESIGNER

Robert Hull, David Miller
Miller/Hull Architects
911 Western Ave., Rm 220
Seattle, Washington 98104

Estimated Building Energy Performance

BUILDING DATA

Heating degree days	6,542 DD/yr.
Floor area	2,700 sq. ft.
Building cost	$180,000
Cost of special energy features	$ 8,000
Total building UA	1,820 Btu/°F. hr.
Heating	
Area of solar-heat collection	450 sq. ft.
Solar heat	90 percent
Auxiliary heat	10 percent (wood)
Cooling	
Natural means	convection, shading and earth tempering
Hot water	
Need	21,000 gallons/yr. @ 120°F. (incoming 60°F.)
Auxiliary electricity cost	$30/yr.

ANNUAL ENERGY SUMMARY

Nature's contributions	
Heating	638.0×10^6 Btu
Cooling	85.0×10^6 Btu
Hot water	10.3×10^6 Btu
Auxiliary purchased	
Electricity	37.0×10^6 Btu or 13,711 Btu/sq. ft.
Fuel wood	not calculated

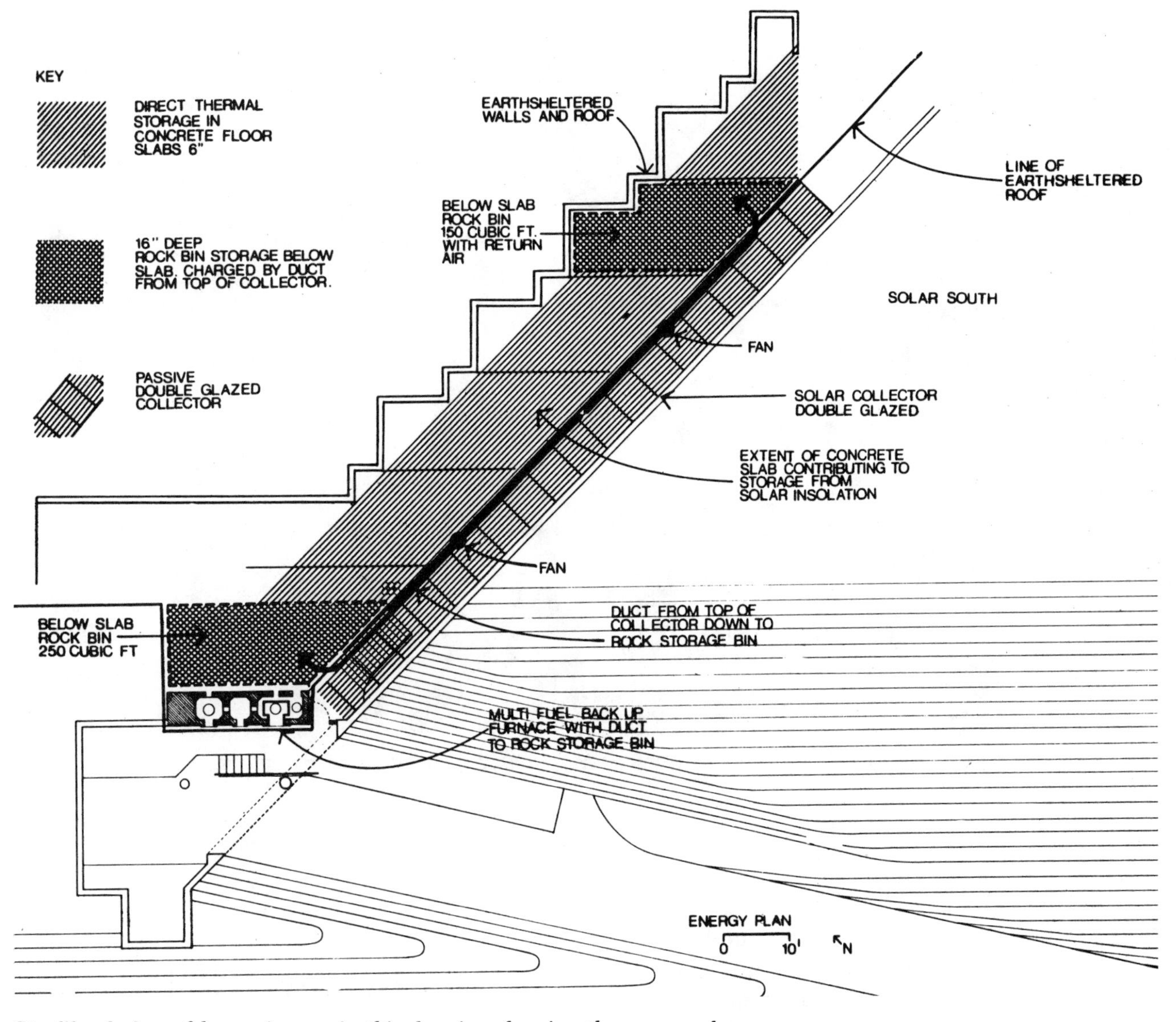

Steplike design of house is seen in this drawing showing the energy plan.

To gain maximum exposure to the sun, architects David Miller and Robert Hull devised an ingenious plan for their client's proposed home in northwest Washington state. They designed the house so it appears to step up diagonally across a steep, south-facing slope in the Cascade Mountains. The steps are a series of bedrooms and studios, connected by a diagonal ramp, and a series of articulated, outdoor terraces. By placing the rising terraces on a diagonal, each 3 feet higher than the next, there are views both down into the lower terraces and out into the valley and mountains. A high-ceiling living room dominates the interior and also provides spectacular, panoramic views.

A Diagonal Solar Collector

The diagonal emphasis grew out of the need to place the solar collector in the most advantageous position. In order to face due south for maximum efficiency, the collector was placed on a diagonal

House fits into steep, south-facing slope of the Cascade Mountains; ramps and terraces make plan possible.

plane across the contours of the mountain. This suggested the rooms behind could be terraced and linked by a ramp. The vectors of the ramp and collector became the diagonal axis for the house. Thus, the house itself became a lineal diagonal element on the steep incline, as a counterpoint to the access road and driveway ramp to the garage at the base of the house. At the intersection of these two diagonals is the lowest terrace, the point of entry and a place for service and thermal storage space. Directly above are the living spaces.

The sloped solar collector is a continuous glass wall facing due south at a tilt of 20 degrees and made of roll-up garage doors turned on their sides so they can roll sideways. The wood door frames are factory fitted with double-glazed panes. They can be opened in full height sections during the summer for ventilation. This inexpensive glazing system is used throughout the house as a standard solar window wall.

A Northwest Lodge

Inside, the clients wanted to recreate the traditional feeling of a northwest lodge with heavy-timber construction. However, because of the precipitous access road, structural members could not be very long. Heavy snow loads on the roof also required relatively short spans. The proposed structural system keeps all the spans to a maximum of 8 feet. To avoid a grid of columns at 8-foot centers, the designers developed a structural tree concept of columns with radiating braces to the roof. A local log-cabin manufacturer will supply the 10-inch round milled log components cut to specifications.

The feeling of a lodge is enhanced by a living room fireplace. The fireplace also provides a secondary heating source, backing up the home's furnace and passive solar system. The nearby kitchen space is raised 3 feet and looks out through the high living room windows. The dining room overlooks the sunset viewing court to the west of the living room. The utility room, which includes a laundry, is large enough to accommodate special seasonal activities, such as preserving and butchering, that might expand out from the kitchen.

Since Barney Kimmick is a painter and Nancy is a photographer, there was a need for two separate but open studios. The photography studio opens to a sunrise terrace that terminates at the upper southeast end of the lodge.

Earth Sheltering

Because of the steepness and severity of the 4,000-foot-high site, it is inevitable that to some degree the house should be dug into the mountain, especially since the earth is workable. Despite the steep, 22-degree slope, the mountainside is not solid rock, but a loose mix of rock and soil. The absence of sheltering trees also reinforces the need to build the house low into the ground to minimize climatic effects. Winter snowfalls of 3 to 4 feet are common; summer temperatures exceed 90°F.

The earth on the roof helps temper the climate and allows the house to be integrated aesthetically with native growth and forms. The native semiarid vegetation is typical upland chaparral, and the scars of reshaping the earth will heal easily.

On the north side, the earth is held back by an insulated, concrete retaining wall, a wall with a corrugated plan that provides structural stiffness and is part of the geometry to reduce roof spans. The corrugation also extends the length of the wall, to add more exposed masonry mass for thermal storage on the interior.

The wall has 2 inches of rigid insulation on the exterior. The perimeter footings are insulated vertically to provide a thermal break between the ground under the floor slab and the earth outside. The roof uses a plywood deck with a continuous membrane of .06-inch-thick sheet rubber. Above that will be 3 inches of water-resistant foam insulation (extruded polystyrene), which will protect the membrance from puncturing by earth fill or tree roots.

Built-in Thermal Storage

The solar heat from the direct-gain window wall is partially stored in the exposed concrete walk ramp and in the end walls of terraces and their slab floors. The concrete slabs are 6 inches thick and etched and stained to enhance their color and to improve their heat-absorption characteristics. There is sufficient exposed concrete to absorb most of the solar direct gain.

A thermostatically-controlled duct at the top of the window wall will collect high-temperature stratified air and direct it by blower into two rock-storage bins. The bins, composed of round river rock, are located under the living room and the east bedroom. The bins are large (18 × 20 × 1.5 feet) and flat, so their heat can be delivered by conduction through the floor slabs above. Placed at both ends of the house, they provide alternate heat sources. The lower rock storage bin can also be charged by heat from a multi-fuel furnace, which would be capable of burning wood, brush and refuse. The furnace, which is accessible from the garage level, is the primary heating backup system.

A Distinctive Mountain Retreat

Throughout the seasons, the ramped terraces of this unusual mountain lodge will provide an unimpeded view of a constantly changing mountain landscape. In winter, this earth-embraced passive solar house will be snug. Its variety of levels and exposures will undoubtably develop a series of distinctive interior micro-climatic responses, depending on the weather. In summer, the house can be opened to varying degrees. In many ways, it can operate as an open earth-sheltered pavilion, exposed to clear mountain air and dramatic views.

Sundance One For Virginia

RESIDENCE FOR

Walter F. Roberts, Jr.
Reston, Virginia

DESIGNER

Walter F. Roberts, Jr.
Box 2730
Reston, Virginia 22090

BUILDER

Walter F. Roberts, Jr.

STATUS

Occupied, 1979.

Estimated Building Energy Performance

BUILDING DATA

Heating degree days	4,962 DD/yr.
Cooling degree days	940 CDD/yr.
Foundation area	784 sq. ft.
Total floor area	2,236 sq. ft.
Building cost	$130,000
Cost of special energy features	$ 9,800
Total building UA (day)	750 Btu/°F. hr.
Total building UA (night)	350 Btu/°F. hr.
Heating	
Area of solar-heat collection	792 sq. ft.
Solar heat	68 percent
Auxiliary heat type	electric baseboard
Auxiliary heat needed	three window-mounted heat pumps
Cost of auxiliary heat (1980–81)	$150
Cooling	
Earth tubes, shading, natural convection	
Auxiliary cooling needed	none (circulation fans used occasionally)
Hot water	
Need	27,500 gallons/yr. @ 140 °F. (incoming 56 °F.)
Electricity cost	$240/yr.

ANNUAL ENERGY SUMMARY

Nature's contributions	
Heating	27.2×10^6 Btu
Cooling	32.3×10^6 Btu
Hot water	planned
Auxiliary purchased	
Electricity	28.0×10^6 Btu or 12,540 Btu/sq. ft.

It is quite natural that a professional known for "energy-responsive architecture" should house his family in an award-winning residence of his own design called "Sundance One." In creating this design, the architect's goal was, "to develop a basic, single-family house which would require a minimum of external energy to maintain a comfortable internal environment."

But, to achieve this objective in Reston, Virginia was not a simple challenge. This community near the Washington, D.C. beltway has grey, damp winters, conditions seemingly inappropriate for building solar. Moreover, the summer has many warm, muggy days.

Walter F. Roberts' architectural solution was to build two houses, one inside the other like a double shell. But, unlike traditional brick houses in the mid-Atlantic states, this design has the masonry shell on the inside, and concrete block walls are the exposed interior materials.

"The living spaces are all contained within the mass of the inner house which moderates temperature swings within those areas," Roberts says. "The areas contained between the two houses act as buffer zones as well as collection and distribution zones for heating and cooling. The outer house is a cocoon encasing and protecting the inner house from climate changes."

This outside house is an insulated wood frame. The architect likens it to a "cocoon" because of its sheltering enclosure of 2×6 construction with movable insulated roll-up curtains behind the single-glazed windows. The value of this concept is that more than 60 percent of the winter heating and almost all of the summer cooling is provided passively.

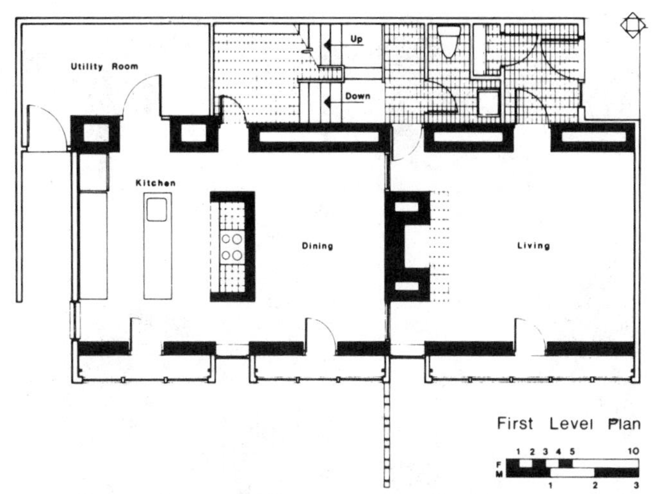

Heavily lined areas indicate masses of masonry in this house.

The Operation of the Cocoon

The house temperature is controlled by opening the cocoon or outer house to allow the penetration of the sun's rays or cool breezes, depending on the season.

"During the winter months, as the sun begins to warm the south face of Sundance, insulated curtains which cover the entire south side of the house open," Robert explains. "The sun then begins to warm the inner-mass house. As the south wall is heated, warm air within the sunspace rises up into the greenhouse/attic spaces. This warm air activates fans located at the top of the north mass wall. The warm air is then driven down through the north mass walls, giving up its heat to these walls. The air then returns through the crawl space to be reheated as it enters the sun space. As darkness falls, so do the thermal curtains, preventing the escape of heat to the outside, and allowing the warmth to radiate into the house. For about 12 hours each day, a small oscillation fan is used in each of the main living areas to assist the transfer of heat from room air to the masonry mass.

"During the summer months, the insulated curtains are closed during the day preventing the sun's rays from penetrating the house and warming the inner mass. The windows are also closed preventing the penetration of hot humid air. With the windows closed the only air which can enter the house is through the earth tubes. The two thermal chimneys located at the roof's peak pull earth-cooled air through these tubes and up within the sun space, cooling the south mass wall and then being exhausted out the chimneys. When the outside air cools in the evening, the curtains open, the windows open and cool air draws through the living spaces up the stairwell to be exhausted by the chim-

Wood sculpture and light featured in main entrance.

neys. This process flushes the inner house of any heat built up during the day."

Looking at Outside Details

Outside, the severe rectangular mass of this three-story house is relieved only by the stepped plan which moves up the hill 3 feet at the midpoint. The pyramidal roof form is capped by three massive silhouettes. The center one contains fireplace flues. The end ones are thermal chimneys for summer ventilation. Even though the south wall is totally glazed for windows or passive solar, heated-air collection, its appearance is modifed by wood trim and the detailing of two recessed window bays. But, when the movable insulation is in place, its reflective foil surface does change the outward appearance.

The Order of Interior Spaces

The interior spaces are ordered by the same classical simplicity and concern for formal clarity as the exterior elevations. Hollow masonry walls conceived as air ducts separate major-use rooms from the circulation and services on the north side. Thus, both the masonry walls on the south side as well as the interior masonry walls are warmed by the sun, and both sides of major rooms are bounded by masonry walls with thermal-storage and radiant-warming capacities. To maximize usable space within the total form of the house, there is a third floor or loft level that provides studio and greenhouse space up under the sloped planes of the roof.

The Craftsman Aesthetic of the Interior

The aesthetic character of restrained tailoring on the exterior is continued on the inside, with very different materials. Full-weight concrete blocks with grouted cells to add to the mass are used consistently throughout. Their natural color comes from gravel aggregates. But, their design quality comes from alternating the block size by course. The distinction comes from the carefully resolved elevations in which every mortar joint is in place within a module of masonry unit dimensions. The color effect of this crisp visual setting is a blend of subdued polychromy which serves as a rich background for oriental rugs and elegant furnishings. Overhead, are exposed, bolted, built-up beams and purlins. Their decorative quality is derived by revealing their careful assembly.

Integration of Passive Components

An aesthetic integration of every material and surface distinguishes the exceptional custom quality of

Masonry and ceiling woodwork are highlights of living room.

Sundance One. Similarly, the passive heating and cooling components are integrated completely with the architectural expression. The insulated curtains that blanket the south wall are completely automated, and, when installed, were among the largest of their type. South windows open into the Trombe wall space and allow solar-heated air to circulate directly into the rooms. Normally, the windows are closed and air thermosiphons up through the attic space and down through the hollow north masonry walls where its heat is delivered indirectly by radiation.

The summer cooling is similarly integrated within the architecture. Since the house has no air conditioning system, it is noteworthy that the cool air movement system together with the thermal and moisture interchange of the masonry walls has kept the interior in the comfortable 70-degree range through several humid Virginia summers. The design is exceptional in that the passive solar heating system does not add to the summer cooling problem but is integrated with the passive cooling system.

Technical Innovation Integrated With Distinct Aesthetic

It is not surprising that this house was unanimously selected for recognition. It has been noted in many other places, both for its technical innovation and for its aesthetic integration of new thermal concepts. Because of the distinct character and clarity of parts, several material suppliers have used the house nationally in published advertisements. The house has also been monitored by the National Solar Data Network. It has been instrumented for both conventional as well as comparative studies using innovative measuring techniques. Thus, in addition to conventional calculated predictions made in 1978 and the actual readings of fuel and electric meters, there are precise records and analyses to confirm its actual performance (see Appendix A).

Evaluating Comfort and Style

Among the few criticisms of the house might be the layout of the entrances. A front door enters the northeast corner of the house from the formal walk that steps down from the driveway. An enclosed entry hall then properly opens directly into the formal living room. However, the service entry to the house is awkward. By placing the kitchen on the southwest corner to maximize views into the most private part of the two-acre wooded lot, there is no obvious way to get groceries in or garbage out without walking through the living room.

Both for character and for comfort, Sundance One appears to be a great success. For some, the severity of the interior may not be to personal tastes. But, the identical materials could be painted or even wallpapered with hardly any thermal effect. Similarly, softer and more decorative furniture would transform what some might regard as an austere interior. But, these are questions of personal taste that would not affect the heating and cooling comfort delivered by natural means.

A Second Home In The Virginia Mountains

RESIDENCE FOR

Dr. and Mrs. Maurice Woods
Wintergreen, Virginia

DESIGNER

Edward Mazria, A.I.A.
Marc Schiff, A.I.A.
Thomas Cain, Job Captain
Mazria/Schiff & Associates, Inc.
Box 4883
Albuquerque, New Mexico 87196

BUILDER

Pat Fincham
State Bridge Construction
Route 1, Box 291
Shipman, Virginia 22971

STATUS

Occupied, Spring 1980

Estimated Building Energy Performance

BUILDING DATA

Heating degree days	5,250 DD/yr.
Floor area	1,500 sq. ft.
Building cost	$80,000
Total building UA	468 Btu/°F. hr.
Heating	
Area of solar-heat collection glazing	540 sq. ft.
Solar heat	77 percent
Auxiliary	heat pump (COP 2.0)
Auxiliary heat needed	8,683 Btu/sq. ft./yr.
Cost of auxiliary heat	12¢/sq. ft./yr.
Cooling	
Natural convection	
No auxiliary necessary	
Hot water	
Need	26,000 gallons/yr. @ 130 °F. (incoming 55 °F.)
Active collectors	75 sq. ft. electricity
Auxiliary cost	$70/yr.

ANNUAL ENERGY SUMMARY

Nature's contributions	
Heating	$72. \times 10^6$ Btu
Cooling	no estimate
Hot water	11.2×16^6 Btu
Auxiliary purchased	
Electricity	24.3×10^6 Btu or 15,677 Btu/sq. ft./yr.

The builder of the Woods house, Pat Fincham, wanted a custom design for a home in Wintergreen, Virginia, about two and a half hour's drive southwest of Washington, D.C. To get it, he sought architects from 2,000 miles away who could meet his special goals: to increase his potential as a home builder by offering an ideal house for the second-home buyer; and to build with passive solar heating, a feature then unique to the neighborhood, and a way to reduce the costs of heating oil and expensive electricity.

The designers, Edward Mazria and Marc Schiff of Abuquerque, New Mexico, met these goals. They did so by designing a house that conformed to the style and price range of neighboring homes in form and in exterior finish materials. As solar architects, they also saw an opportunity to design a house that effectively used the local site microclimate for heating and cooling, while responding to the client's needs and budget.

In the Woods' house, the architects demonstrated that a well-conceived, passive solar residence can be constructed at no additional cost per square foot than a conventional residence. The Woods' house also shows that energy economy can be built into an attractive house without ostentation. The widespread availability of such reasonably priced solar houses would mean considerable savings both to the owners and to the community.

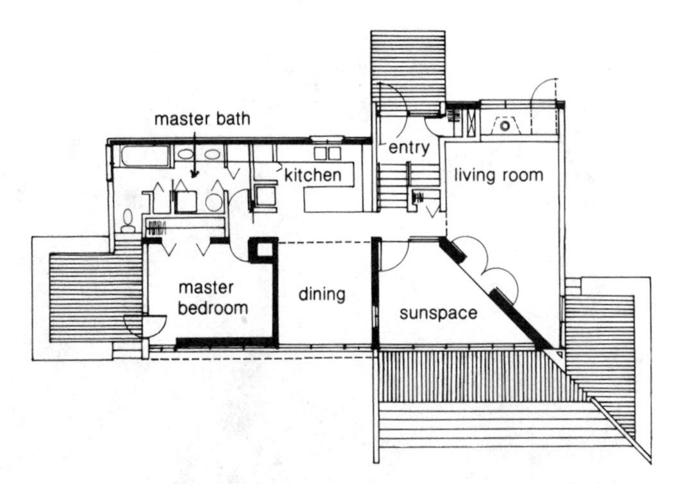

Two-story sunspace provides beauty, passive solar heat, and an ideal location for a display of flowers.

The Designers' Economic and Energy Goals

In designing this Virginia home, Mazria and Schiff applied the same ethic that characterizes all their work:

"Energy efficiency by passive means was a major determinant in the development of the design. Orientations, the size of the openings, groupings of rooms, form of the building and construction materials were all considered, selected and designed to provide the maximum amount of passive heating and cooling while consuming a minimum amount of conventional energy sources."

Within their practice, the Woods' house was one of a series of explorations to clarify passive solar house design for the merchant/builder. The intent was a relatively standardized house design of contemporary but unexotic appearance. Within that standard, they could extend dimensions to suit client, climate and builder. The clarity and simplicity of the Woods' house illustrates the logic of such a house, which avoids looking like either a regional style or a custom-designed piece.

A Mountain Site with Views

The Woods' residence is on a heavily wooded, south-facing slope of a mountainous Virginia ski resort. It is located near the top of a ridge, the land sloping way to reveal views of the valley and surrounding mountains. Though forested, the site is exposed to winter winds. The trees near the house were thinned and pruned to give solar access, but the dense leafy vegetation covering the rest of the site was left for protection and appearance. Both the dense foliage and the elevation, 4,000 feet, make the site cool in summer.

The road approaches the house from the north. Cars are parked near the road, and visitors reach the house by walking through the trees. The entrance is at a half-level deck that permits the main floor to be partially buried on the north side.

Informal path leads to entry. This entry offers easy access to living room, sunspace, dining room, and kitchen.

The east-west axis of the house exposes a long south-facing side for maximum winter solar gain. The natural contours of the site allow the north side of the house to be dug into the slope, limiting winter heat loss and diverting winter winds up over the house and protecting the outdoor space along the south side.

South-Oriented Traditional Living Spaces

The traditional living spaces in the house are designed to conform to the prospective owner's expectations. For privacy, the master bedroom is located on the lower level, with the second bedroom and loft for children or guests placed on the second level. The living areas of the house are on the south side, so the owners can take advantage of solar heat gains and enjoy the views. Less frequently used areas are located along the colder north side where they act as buffers.

One of the most pleasant south-facing rooms is a generous skylit sunspace that opens toward the southeast and extends outward to a roofless porch. The sunspace windows maximize heat gain and frame the distant views.

Passive Heating with Interior Thermal Mass

Passive solar heating and cooling methods are integral to the building design. An unvented, masonry, thermal-storage wall provides nighttime heating in the master bedroom, allowing floor carpeting to be used without compromising the system. This solar wall, also called a stagnating Trombe wall, has no movable insulation. Windows in other living spaces allow direct sunlight to heat these spaces. Daytime solar-heat gain is stored in the thermal mass of the walls and floor of each space for nighttime heating.

The sunspace collects and distributes heat through masonry common walls to the dining and living rooms. The interior thermal mass consists of 8- and 12-inch concrete-block walls filled with a dense slurry; the unit weight is 140 pounds per square foot. The concrete floor slabs have a 1-inch soapstone finish. The second floor has a 2-inch setting bed of grout with a 1-inch soapstone finish over a light frame substructure.

When excess heat accumulates in the sunspace, it is drawn off with a fan and air-duct system and stored in a rock bed under the north end of the living room floor. The heat from the rock bed is distributed passively by direct contact between the rock bed and the living room concrete floor slab. The control system is of the simplest sort, without any temperature sensors in the rock bed itself. There is a cooling thermostat high in the sunspace, and there is a manual override switch. The logic is that the rock bed can always store more heat when air temperatures in the sunspace rise above 85 °F.

The passive solar elements of the building (when continuously occupied) supply 77 percent of the heating needs. Back-up heat is furnished by a controlled-burn fireplace and by electric baseboard heaters. Domestic hot water is supplied by 75 square feet of active solar collectors.

Thermal Mass for Cooling

The house is cooled in summer by nighttime ventilation of the same distributed thermal mass. Cool night air moves throughout the house by natural convection through operable windows on both levels and high vents in the sunspace and the stair hall. Surrounding trees also shade the exterior envelope on the east, west and north sides to minimize solar-heat gain.

Construction for Economy and Comfort

To keep the building costs in line with standard construction, the concentrated thermal mass in the building interior allowed the envelope to be constructed of less expensive wood frame. The exterior walls have 2 × 6 studs with tongue-and-groove eastern cedar siding—familiar and available materials. Six inches of fiberglass insulation within the plasterboard drywall give an insulating value of R-19. The concrete foundation has 2-inch rigid perimeter insulation; its value is R-8.4. The roof is wood frame, with 9-inch, foil-faced batt insulation (R-29), ¾-inch plywood decking, asphalt shingles and sheetrock for the ceiling.

All windows are double glazed in natural-wood frames. Although the interior trim was detailed to accept movable, roll-down insulation, it has not been considered necessary.

The thermal performance of the house over several seasons has shown that it maintains stable temperatures when unoccupied. For a second home, this means no danger of freezing pipes; instead, the house is comfortable and ready to occupy. But beyond the economy of this thermal efficiency is a pleasant relaxed environment completely in tune with its sylvan setting.

Door in sunspace opens onto roomy deck area.

A Playful House In The California Foothills

RESIDENCE FOR

David Brodhead
La Honda, California

DESIGNER

Richard Fernau and Laura Hartman
1555 La Vereda
Berkeley, California 94708

CONTRIBUTORS

David Brodhead, General Contractor, La Honda, California
Walter Waik, La Honda, California
Norm Nelson, Masonry Contractor, La Honda, California

BUILDER

Charles Davis, Design
James Axley, Thermal Analysis
Phillip Ceasar, Domestic Hot water
Raymond Lindahl, Structural
Fred and Ted Jacobs, Mechanical
Bruce Corson, Site Planning

STATUS

Occupied, December 1980

Estimated Annual Building Energy Performance

BUILDING DATA

Heating degree days	3,550 DD/yr.
Cooling degree days	90 CDD/yr.
Total area	1,560 sq. ft.
Building cost	$100,000
Cost of special energy features	$20,000
Total building UA	594 Btu/°F. hr.
Heating	
Area of solar-heat collection glazing	459 sq. ft.
Solar heat	70 percent/yr.
Auxiliary heat	electric resistance
Auxiliary heat needed	13,000 Btu/sq. ft.
Cost of auxiliary heat	27¢/sq. ft./yr.

Cooling

Effective shading, noctural ventilation and thermal mass
No auxiliary necessary

Hot water	
Need	21,900 gallons @ 130°F. (incoming 60°F.)
Auxiliary electricity	$47/yr.

ENERGY SUMMARY

Nature's contributions	
Heating	47.3×10^6 Btu
Cooling	not available
Hot water	10.4×10^6 Btu
Auxiliary purchased	
Electricity	22.6×10^6 or 14,500 Btu/sq. ft. (fireplace wood not included)

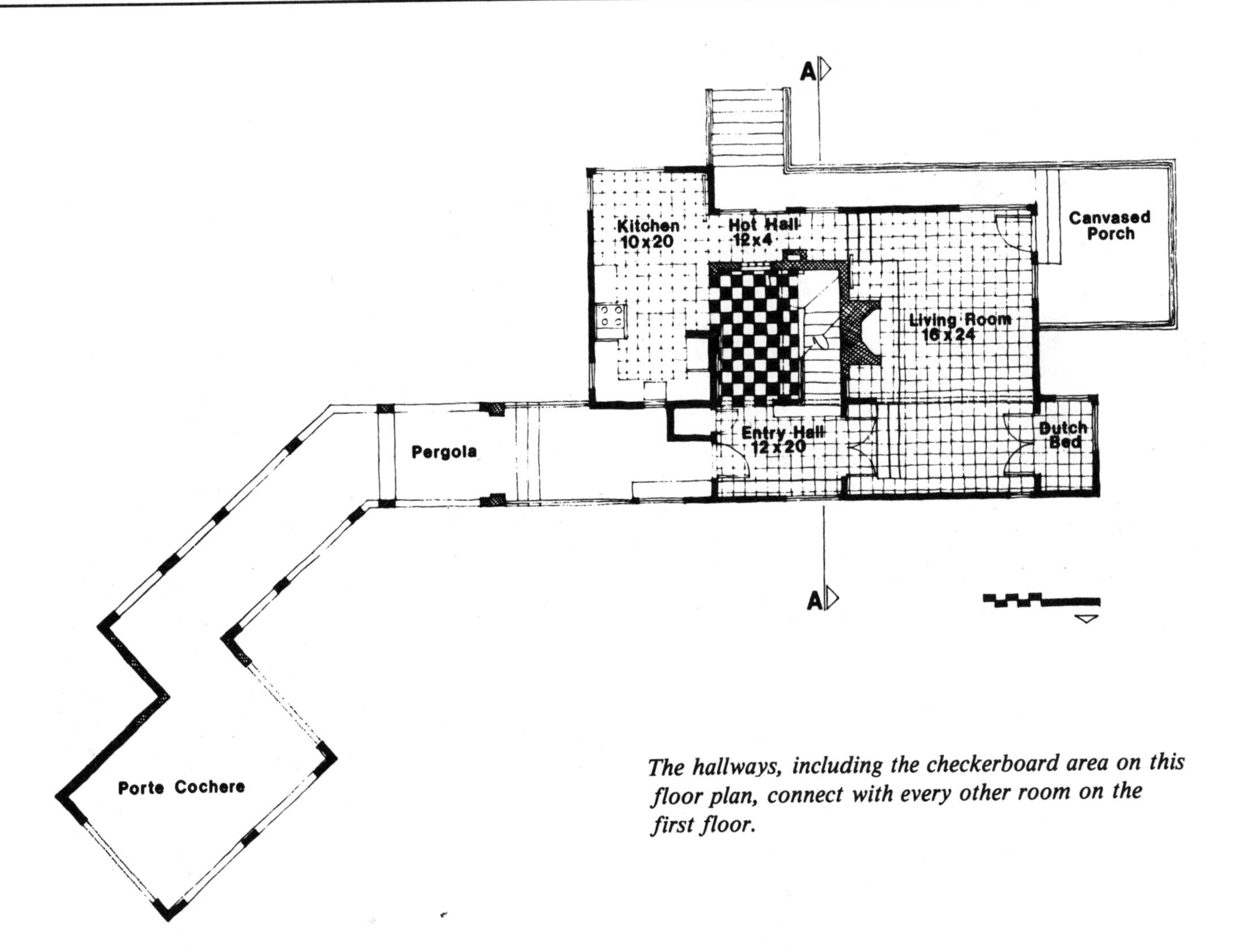

The hallways, including the checkerboard area on this floor plan, connect with every other room on the first floor.

The Brodhead house is located in La Honda, a rural/suburban community in the California coastal range 60 miles south of San Francisco. It is situated near the ridge of a heavily-wooded northeast slope; at this altitude, morning fog is common. The house faces to the southwest, the direction of winter storms. In the summer, there are cooling breezes from the north. This temperate coastal climate requires both heating and cooling.

The client, David Brodhead, is fond of simple vernacular achitecture, and he wanted his house to be sympathetic to its rural surroundings. He needed a modest two-bedroom house. Although intended for year-round use, the house was to be small and use passive-solar heating and cooling. A masonry fireplace was to be included, since firewood could be obtained easily on the 30-acre site.

Starting with Formal Design Choices

Designers Richard Fernau and Laura Hartman made two formal choices at the beginning of the design. The first was to start with a simple gable roof on a 1½-story wooden box that could be "remodeled" during the design process. This choice was based on the need for economy and their interest in recording the "makeshift" adaptations that would result from the impact of the thermal criteria on this generic building form.

The second was to divide the house into thermally distinct rooms. This choice would allow many possibilities for control of overheating, un-

The south elevation, with central entryway.

derheating, natural convection, and natural ventilation.

An Unusual Calculation Procedure

Although most American designers use variations of design methods derived from passive performance research at the Los Alamos National Laboratory, Fernau and Hartman used concepts from the British building thermodynamic analysis method, "The Admittance Procedure," during the formative period of design. This approach helped them in using a building-as-system approach. Later, a computer program, ADMIT, based on this British method, was used to analyze the response of the building to a critical winter day and a critical summer day, and thus fine-tune the design.

Organizing Around a Central Hall

The two-story house that emerged reveals the influence of the initial choices, as confirmed by the thermal analyses. There is nothing unusual about the floor plan except the insistence on separate and discrete rooms for separate functions. In plan, the house is composed of rooms organized around a central entry hall. This grand hall functions as a thermal buffer and as a vertical heating and cooling riser as well as a circulation core. All other rooms are connected to it, either by doors or interior windows that function as dampers.

Architecturally, the entry hall is treated like the outside of the house. Though it is totally enclosed, exterior siding and double-hung windows underscore the architecturally ambiguous character of the room.

A Living Space and a Passive System

The Brodhead house uses direct-gain, passive solar heating during the heating season, passive nighttime cooling during the cooling season and natural convection and ventilation throughout the year to maintain comfort at minimal energy expense.

The house was designed as a passive system composed of a set of thermally distinct rooms, each of which is, to a degree, both a system component and a living space. For example, the "hot hall" that connects the kitchen to the living room acts principally as a solar collector yet is also a hall—a living space that may reasonably be allowed to float above and below normal comfort ranges. It is somewhat like a Trombe wall system, yet without its limitations.

The living room, on the other hand, is principally a living space, yet it contributes to the passive performance of the building as a whole. Its south-facing windows and massive tile floor function as a modified direct-gain space.

The subterranean basement may tend to be overly cool, and can be used to condition living spaces during hot periods. This conditioning is achieved by both natural convection and forced convection.

The Masonry in the Central Hall

Central to this building-as-system concept is a massive and extensive masonry wall that wraps around the central entry hall. Decorative surfaces in the entry hall are patterned in an intricate polychromy, more popular in another century. Masonry also faces the fireplace and chimney walls. With the tile floor, these surfaces provide thermal mass directly

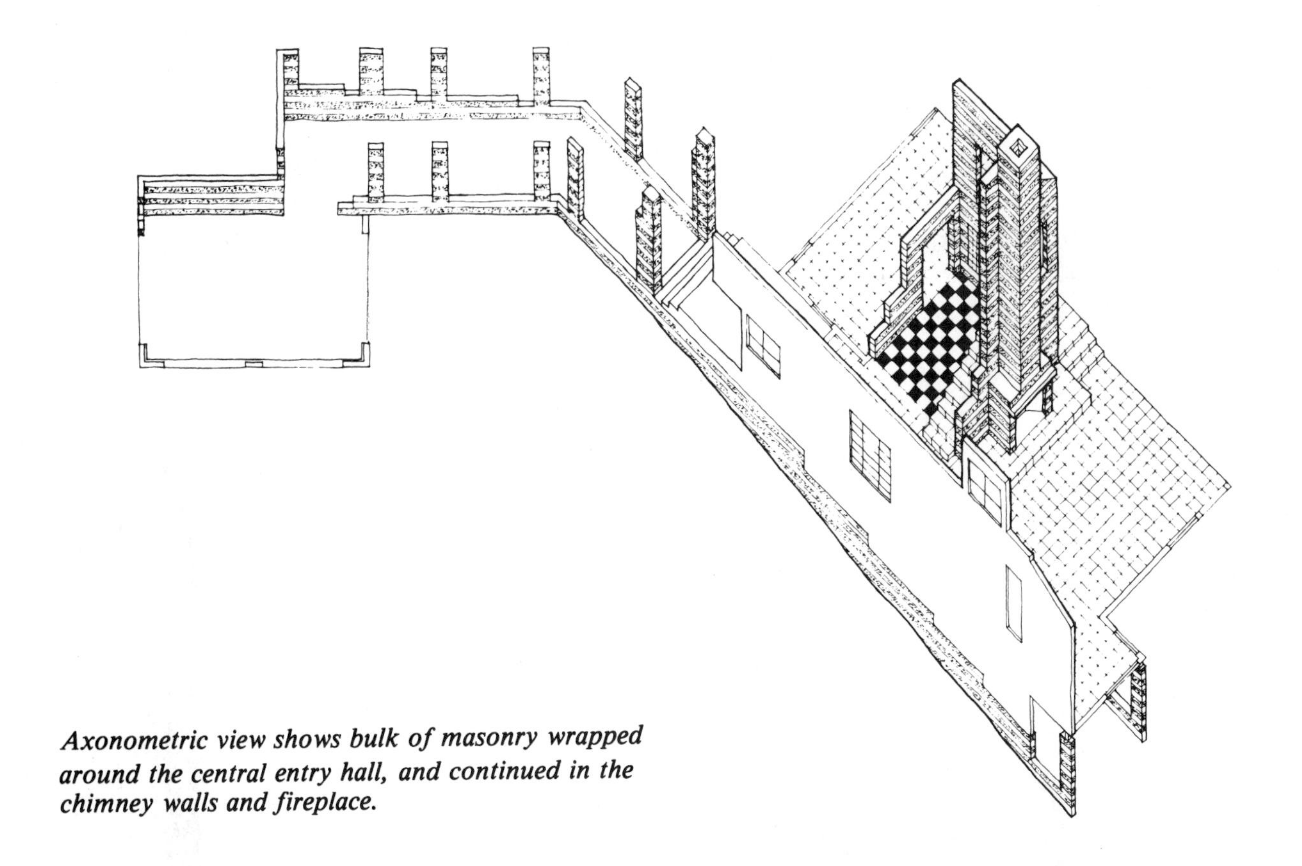

Axonometric view shows bulk of masonry wrapped around the central entry hall, and continued in the chimney walls and fireplace.

to the principal direct-gain collector spaces—the "hot hall" and "warm hall" directly above—and the fireplace. Augmented by the extensive thermal mass of the tile floor, the masonry moderates the intermittent solar heating during winter months and aids night cooling during hot periods.

Moreover, the masonry in the belvedere stores the heat that accumulates there until it is retrieved. In winter, this warm air is recycled to the "hot hall"; in summer, the warm air is expelled by means of the gravity ventilator.

Thermal Analyses Predict House Performance

On the critical winter design day (overcast with 36°F. lows), response analysis indicated that 69 percent of the energy needs could be supplied by direct gain, although a 20,000 Btu/hr. auxiliary source would be needed to keep the house comfortable. The designers believed that an optimally-designed fireplace, with combustion air supplied at the hearth and radiant surfaces maximized, could meet this need. In addition, electric baseboard heating was installed as a convenient back-up.

A Historical Throwback or Something Else?

The Brodhead house has been recognized by several awards and publications. Among the technical details that have been noted are the importance of the belvedere for convective cooling and the Rumford-type fireplace for radiant heating.

Interest in the design, typically, is for its personable architectural character. Even with its compartmentalized rooms, the house is spatially interesting and complex. Some have said the house is gracious and inventive; others remark on its wit or its allusions to regional Bay Area landmarks or to the central planning of East Coast Georgian hall houses. Because it is not a California "see-through" house, many assume it must be some type of "throwback."

A Modest House In A Gentle Climate

PROJECT FOR

Sharon and Bruce Miller
Aptos, Santa Cruz County, California

DESIGNER

Swatt & Stein
Architects and Planners
5422 College Ave.
Oakland, California 94618

STATUS

Unbuilt

CONTRIBUTORS

Kenzo Handa, Robert Swatt, Graphics
5422 College Ave.
Oakland, California 94618

Eli Noar, Energy Consultant
1810 Euclid Ave.
Berkeley, California 94701

Estimated Building Energy Performance

BUILDING DATA

Heating degree days	2,900 DD/yr.
Cooling degree days	25 CDD/yr.
Floor area	1,245 sq. ft.
Building cost	$60,000
Cost of special energy features	$ 6,000
Total building UA	491 Btu/°F. hr.
Heating	
Area of solar-heat collection glazing	104 sq. ft.
Solar heat	70 percent
Auxiliary systems	gas furnace plus wood fireplace
Auxiliary heat needed	10,112 Btu/sq. ft./yr.
Cost of auxiliary heat	2¢/sq. ft./yr.
Cooling	
Natural convection and movable shading	
No auxiliary needed	
Hot water	
Need	25,500 gallons/yr. @ 135°F. (incoming 60°F.)
Auxiliary electricity cost	$26/yr.

ANNUAL ENERGY SUMMARY

Nature's contributions	
Heating	18.9×10^6 Btu
Cooling	no estimate
Hot water	20.1×10^6 Btu
Auxiliary purchased	
Electricity	15,490 Btu/sq. ft.

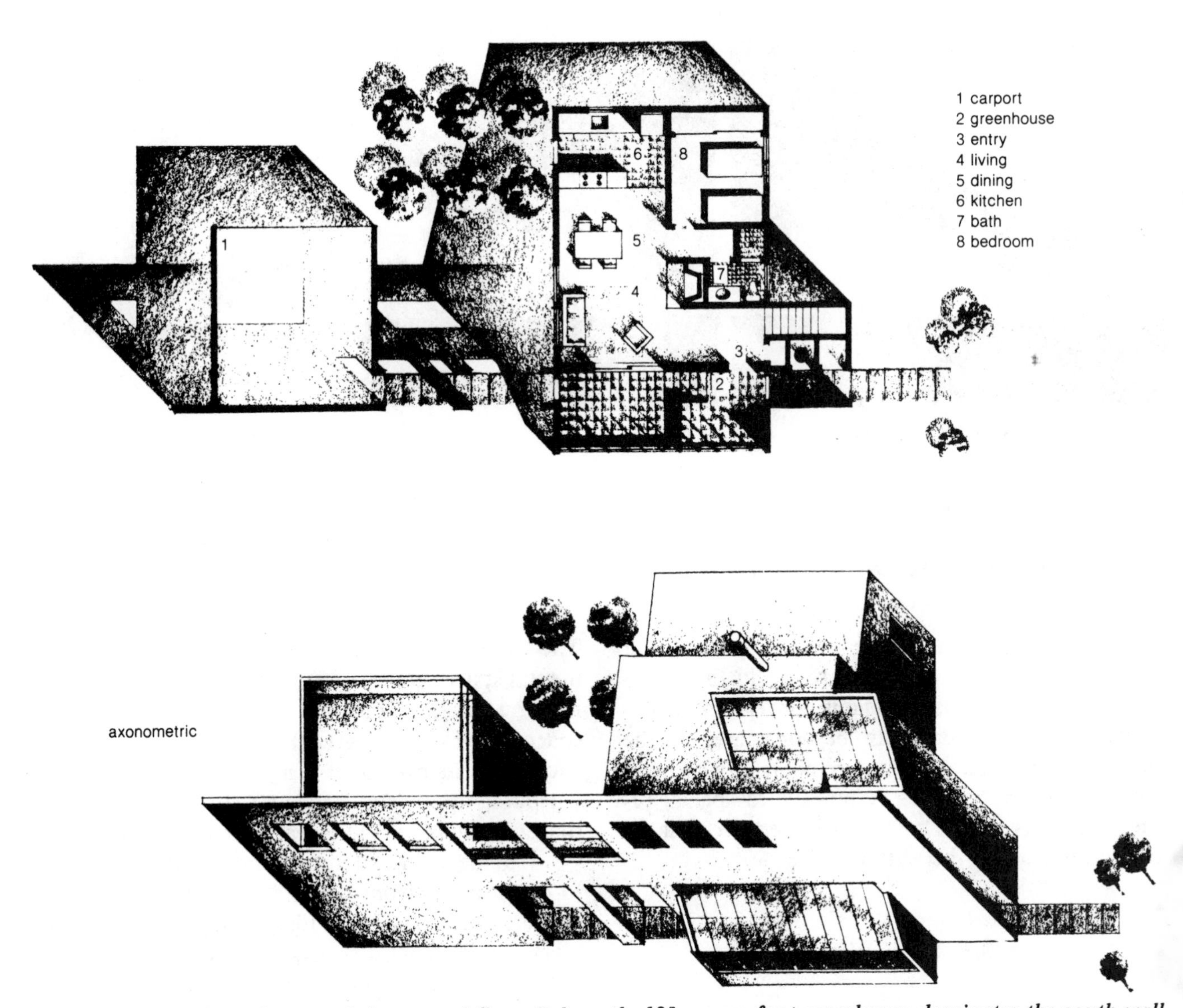

Top illustration shows layout of the ground floor. Below, the 185-square-foot greenhouse dominates the south wall.

This modest residence for a young couple and their child has been designed to meet three specific requirements: first, the construction budget of $60,000 profoundly affected the size, materials and construction techniques. Second, the house needed to be highly livable within the limitations of a 1,250-square-foot area. Third, county ordinances required the house to have a minimum of 65 percent solar-assisted heating.

The double-shed enclosure of the house is oriented to the south to take advantage of the magnificent view of the Monterey Peninsula. Planning has provided maximum access to views and cooling breezes at this unique site overlooking the rolling foothills of the coastal range south of San Francisco.

The house benefits from a bold aesthetic initiative. The long south wall becomes a visual spline and a major framing device, with its carefully proportioned openings. While the budget required a small house and economical construction, there was no limit placed on creative solution. As a first house for a young couple, this design would be memorable in any location.

West (left) and south elevations.

Matching Cost with Aesthetics

The challenge faced by the designers, and their goals in meeting it, is best expressed in their own words: "This house reflects our concern that solutions to one problem must be considered in the light of the many varied requirements of a project. Energy systems must be responsive to spacial requirements as well as cost. It is imperative that architecture express this synthesis."

A Long South Wall

The south-facing wall is 23 feet high and 75 feet long. This heat-receiving wall connects the house with the carport and forms the south edge of the courtyard space between the two structures.

Material and color variations have been used to emphasize the meaning of the wall as a vertical plane between the house and the greenhouse. The wall will be clad in Douglas Fir T1-11 plywood, painted in pastel shades of pink to symbolize the heat gathering function of the greenhouse. The stair tower, topped with solar collectors for domestic hot water, will be similarly clad in plywood and painted a pastel shade of blue-gray. The living spaces will have natural cedar shingles to poetically symbolize "home."

Organizing Fundamental Space

The floor plan is a simple organization of fundamental space. On the ground floor, the living, dining and kitchen areas are part of the same open space that faces the greenhouse. The child's bedroom with its own bath is in the northeast corner. Upstairs, the master bedroom suite has windows that look out over the greenhouse to distant views.

The design is interesting in that the approach to the house is from the west. The double carport shelters the house and adds privacy. Thus the primary outdoor patio is between the two buildings, allowing the house to fit easily into its long and narrow site.

The ground floor is concrete slab on grade without basement. Insulation is R-19 foil-faced fiberglass batts in the walls, R-30, foil-faced fiberglass batts in the roof; and 1½ inches thick × 24 inches wide rigid insulation at the perimeter of the floor slab.

Using Solar Energy Simply

The solar heating for this heavily-insulated residence is provided by a combination of direct gain and passive thermal storage systems. To avoid complex ductwork and air-circulating fans, the de-

East (left) and north elevations.

signers placed the living spaces next to the heat collection and storage areas. The two-story loft space and sloping ceilings help to keep air moving between the house proper and the greenhouse.

The 185-square-foot greenhouse is the primary solar collector. Thermal storage of the greenhouse heat is in a 16-inch-thick concrete block wall, which separates the greenhouse from the living room.

The passive heating system has a day mode and a night mode. During the day, the family can open standard registers near the top of the greenhouse to admit warm air to the master bedroom and the upper floor. As it cools, the air travels by convection down through the two-story space at the north end of the upper level and returns to the greenhouse through the sliding glass doors of the living room.

At night, the family can close the vents. The house is then heated by a combination of low-grade radiation from the thermal storage wall and an energy-efficient *Majestic* fireplace stove.

Passive Cooling, Too

The greenhouse and the thermal-mass wall are equally effective in reducing temperature levels in the house during the summer. While the local climate rarely brings conditions above the comfort zone, the greenhouse could overheat the house. Here again, a two-phase strategy is used. During the day, the family extends a heat-reflecting mylar film over the exterior of the greenhouse that allows only a small temperature rise which is sufficient to warm the air and cause it to rise through the upstairs registers and exit through the bedroom windows. On very warm days, the vents to the upstairs bedroom are closed and warm air is expelled to the outside through the greenhouse vents. This venturi effect and natural cross-ventilation provide ample cooling on this breezy site.

At night, the owners can lift the film and allow the greenhouse temperature to drop, also cooling the block wall, which absorbs heat from the house by conduction and as a radiative heat sink. In both summer and winter, the well-insulated walls and roof reduce heating and cooling loads.

Adapting the Design to Other Sites

Although this house design was done for the specific site and the needs of the Miller family, the designers have also produced drawings that show how such a house could be used in a typical development on an infill site, such as a standard 50- × 100-foot lot between two conventional houses facing west. In a second hypothetical study, they suggested a planned subdivision where the design could benefit from the economies of serial construction as a neighborhood development.

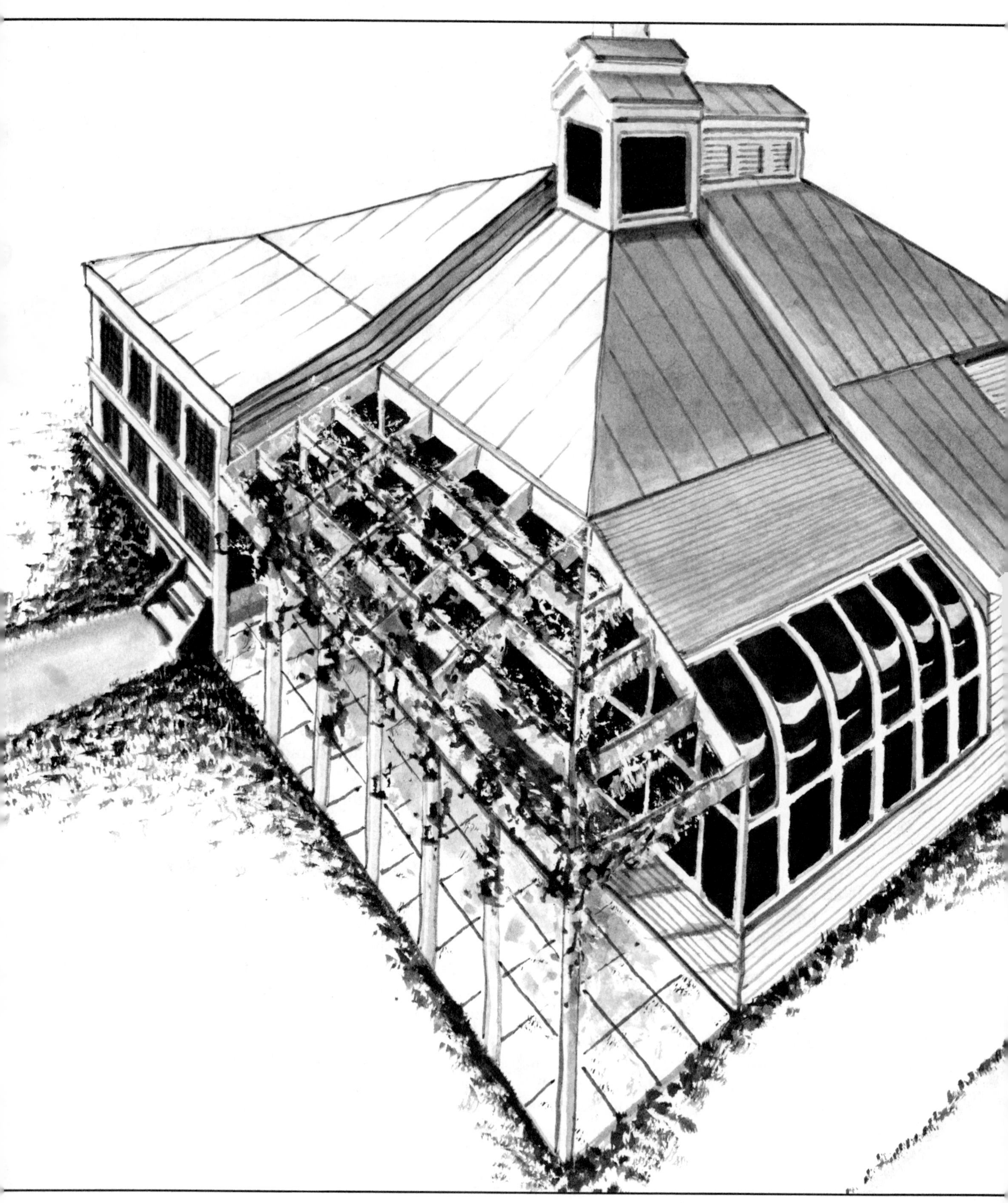

Passive Cooling For The Sunbelt

PROJECT FOR

Passive-Cooling Concept House
Charleston, South Carolina

STATUS

Unbuilt

DESIGNER

Michael Funderburk
Sunshelter Design
1209 Hillsborough St.
Raleigh, North Carolina 27603

Estimated Building Energy Performance

BUILDING DATA

Heating degree days	2,146 DD/yr.
Cooling degree days	2,078 CDD/yr.
Net conditioned area	1,500 sq. ft.
Building cost	$75,000
Cost of special energy features	$ 1,000
Total building summer gain	25.7 Btu/hr./sq. ft.
Total building UA	507 Btu/ °F. hr.
Heating	
Area of solar-heat collection glazings	595 sq. ft.
Auxiliary heat type	electric resistance
Auxiliary heat needed	5,800 Btu/sq. ft./yr.
Cost of auxiliary heat	7¢/sq. ft./yr.
Cooling	
Natural cross ventilation; stack ventilation through roof monitors; induced ventilation	
Auxiliary cooling type	electric fans
Auxiliary cooling needed	4,500 Btu/sq. ft./yr.
Cost of auxiliary cooling	5¢/sq. ft./yr.
Hot water	
Need	16,200 gallons/yr. @ 125 °F. (incoming 60 °F.)
Auxiliary electricity cost	$180/yr.

ANNUAL ENERGY SUMMARY

Nature's contributions	
Heating	16.9×10^6 Btu
Cooling	not calculated
Hot water	not calculated
Auxiliary purchased	
Electricity	15.4×10^6 Btu or 10,296 Btu/sq. ft.

While most North American homes require heating to maintain comfortable inside temperatures, many, especially those in the sunbelt, also require cooling. In Charleston, South Carolina, for example, these two requirements are almost equal: about 2,000 degree days of space conditioning are required in each heating season and in each cooling season. For warm, humid climates, shading and ventilation are crucial to reduce summer overheating. In this passively cooled southern home, the roof design and surrounding vegetation enhance shading. And, at the same time, any side of the house may be opened to take best advantage of prevailing breezes.

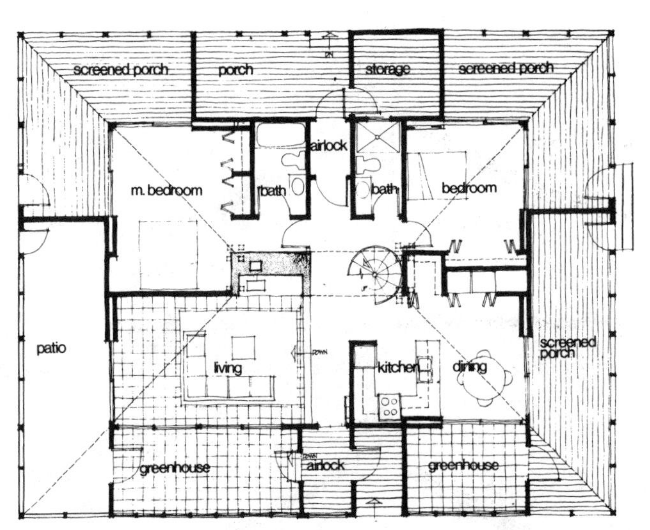

Airlocks at front and rear entrances control a central breezeway.

A Familiar House Plan

Both the plan and the appearance of this house may look familiar. The plan is typical of many simple American two-bedroom builder houses. The front of the house is devoted to living and dining. The central location of the kitchen makes it easily accessible from anywhere in the house.

The two bedrooms have corner locations for cross ventilation. In typical two-bedroom homes, the central bathrooms often share a common plumbing wall, but here they are separated by an airlock. The airlocks at both front and back, in fact, control an enclosed breezeway similar to the through hall of traditional southern houses.

A Traditional Exterior for Warm Humid Climates

Perhaps even more traditional is the exterior appearance. It resembles a vernacular typical of houses built in warm, humid climates. The raised house benefits from breezes, while capturing cool air below. The raised floor is another advantage; it's well above damp soil and pests. Structural pilings also provide economical flood protection. The pyramidal roof is structurally stable against tropical storms such as hurricanes. The broad, hip roof shelters porches on all sides and leads up to a central, vented ridge. The ridge is topped by a ventilation monitor, sometimes known as a *belvedere*. The ridged, sheet-metal roofing is often chosen for buildings in warm, humid climates because of its long life, resistance to growing things, and its high reflectivity and emissivity of solar heat. Solar heat is stopped at the roof before it enters the living spaces and the corrugations help dissipate heat.

In this design, all features—the raised floor, the sheltered porches, the belvedere and the metal roof—are detailed to maximize both their cooling potential and their thermal control. Each feature can help minimize solar heat gain and can assist other parts of the house to lose heat by encouraging natural ventilation. Thermal control is achieved by movable elements so that the house can physically change its appearance through the heat of the day or from season to season. Thus, there is a rhythm of changes in both functional uses and of movable elements based on thermal response to comfort. In this climate, high levels of insulation are not necessary or economical.

A Simple House and A Sympathetic Setting

This is a simple house despite the construction finesse required in the details such as the closable

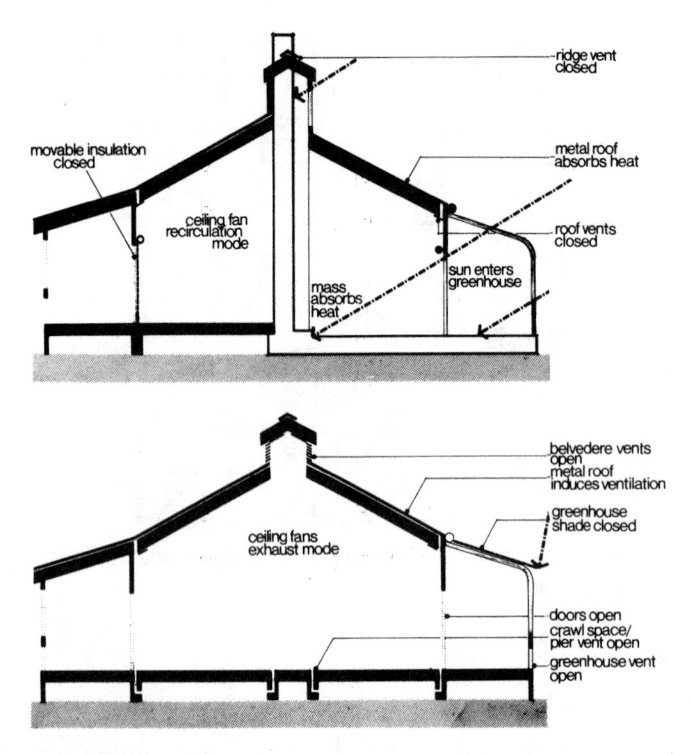

Winter (top) and summer modes illustrate thermal control possible.

belvedere. Its elementary plan would be adaptable to many households, and its simple moving parts allow it to respond to a variety of climatic and seasonal situations. Part of the successful operation of this design depends upon the maintenance of a pleasant tree-shaded garden on all sides. The general modesty of this house and garden design is without any illusion except a gentle acceptance of a surrounding natural setting. Its dependence on daily thermal management assumes a sympathetic and understanding interaction between homeowner and weather, between man and nature.

Concepts for Passive Cooling

This dynamic house has several passive cooling features that require day-to-day attention of the owner for maximum effectiveness. For example, in the morning as the sun comes up the eastern screened porches are closed from the house by drawing the movable insulation curtains. Then south porches are closed, followed by the west and north. Glazed areas on the east and west sides are protected from solar gain by the adjacent porches, and, in the case of the living room, by vine trellises and folding shutters. Solar gain into the greenhouses is blocked by exterior, roll-down, reflective shades.

During the day, ventilation air enters through floor vents at the piers from the crawl space, providing the house with fairly cool air. The ceiling fans can be used to recirculate air inside the house within individual rooms. Air can be exhausted out the belvedere/ridge vent via the central fan. Warm air moving upward to the upper-vent area is assisted by the negative pressure created by wind moving through the belvedere. In addition, the shape of the sloped ceiling that follows the form of the roof acts as a funnel to rising warm air, directing it toward the belvedere. This natural process can be reinforced by the use of ceiling fans in their reverse mode.

As the sun moves to the southwest, the upper part of the central masonry chimney receives solar heat gain through south- and west-facing clerestory windows. The warming of this mass helps induce a draft through the ridge vent and belvedere vents. This draft will strengthen as it maintains sun exposure, peaking in the evening hours, due to the time lag of the masonry.

Natural Ventilation After Dark

In the early evening just after dark, some of the openings between the house and porch can be opened for additional ventilation, if necessary. As the outside air cools down, most or all of the openings between rooms and porches can be opened to induce cool outside air into the house, particularly into the kitchen, living and dining rooms. Heat stored in the thermal chimney will continue to produce a draft to facilitate the ongoing stack effect of exhausting heated air at the top of the house. Also the room ceiling fans and central fans can now be used in the exhaust mode by increasing their speed and by changing their direction.

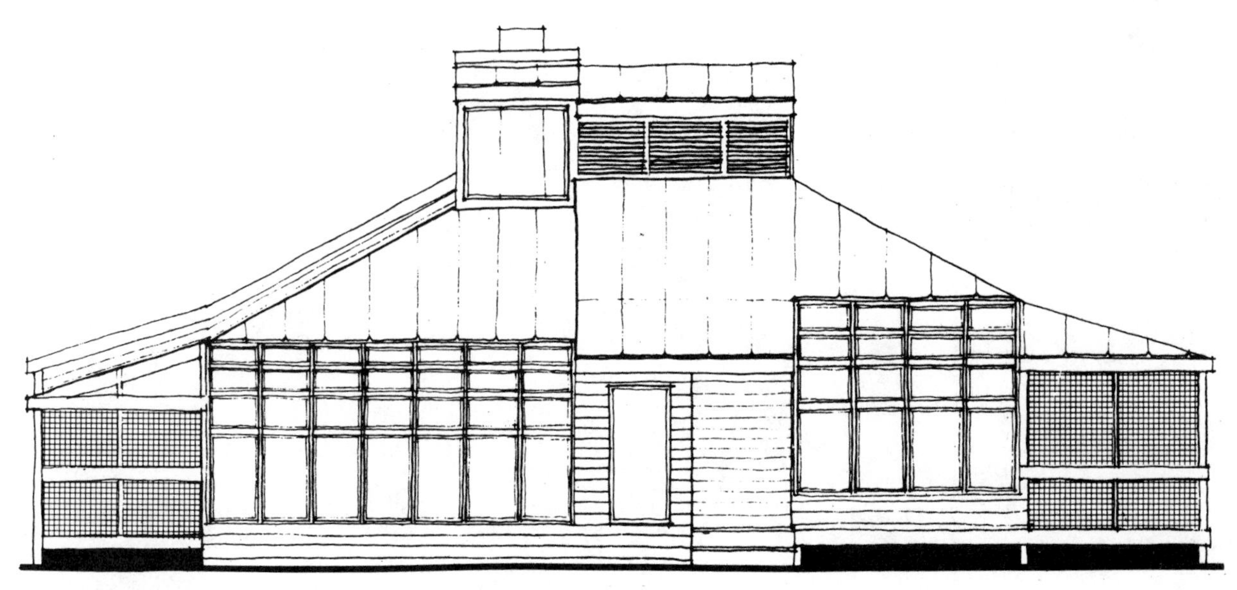

South side. Window at top left is for solar-induced venting.

Ventilation Through the Roof Construction

Another aspect of the ventilation system is air induction through the air space between the insulation and the metal roof. As the sun heats the roof during the day, air enters this space through interior ceilings vents along the perimeter of the building and travels through the roof to the belvedere to be exhausted.

This air space also serves as the roof/insulation ventilation area in the winter time. For this purpose, the interior perimeter ceiling vents would be closed, and the openings to the belvedere would also be dampered so that this ventilation would occur typically from soffit to ridge. By locating half-ridge vents around the base of the belvedere, at its junction with the main roof, the belvedere venting system would be kept separate from the roof-venting system.

Different combinations of inlet systems (pier vents, windows and wall vents) and exhaust systems (cross flow, belvedere/ridge, roof induction) can be used to achieve a comfortable condition by several means: by reducing temperatures sensible heat stress is reduced; by increasing air motion the effect of humidity and latent heat stress is reduced; and by using reflective materials outside and low emissivity surfaces inside, the effects of radiant heat are reduced. Through the movable features and a controlled quantity of thermal mass there is minimal heat gain by day and maximum heat transfer by night.

Passive natural cooling is a more difficult design strategy than passive solar heating. Generally, the effects of cool breezes are much more subtle than the heating capacity of sun flooding through a window. Here, the passive cooling advantage of air motion includes cross ventilation, stack effects and induced ventilation. However, a house open in all directions to maximize the natural ventilation potential is also open to insects, rain and outside dirt as well as loss of privacy. These problems can be addressed by good design.

Here, a concept house design for a warm, damp climate shows how ventilation cooling can be enhanced by thinking three dimensionally. In addition, the orderly organization of the house and its surroundings encourages easy operation and maintenance. Simplicity and symmetry have resulted in a handsome and economical small house.

Heating And Cooling With A Roof Pond

RESIDENCE FOR

Franklin and Nina Fry
Yuma, Arizona

DESIGNER

Guilford A. Rand
Solar Oriented Architecture
16 South 2120 West
Provo, Utah 84601

BUILDER

Joe Schiele
Contemporary Homes
2620 Barbara Avenue
Yuma, Arizona 85364

STATUS

Occupied, June 1978

CONSULTANTS

Energy Roof (Basic Patent No. 3,994,278)
A. Lincoln Pittinger
Consulting Engineer
62 Colonia Mira Monte
Scottsdale, Arizona 85253

Shelter-Rite
Box 331
Millersburg, Ohio 44654

Lynn Vanlandingham and Associates
1832 East Osborn Road
Phoenix, Arizona 85016

Patricia O'Leary Keating
Design Consultant
Miramonte, Arkansas 72705

Actual Average Building Energy Performance

BUILDING DATA

Heating degree days	974 DD/yr.
Cooling degree days	4,125 CDD/yr.
Floor area	1,500 sq. ft.
Building cost	$42,000
Cost of special energy features	$10,200
(includes auxiliary heat pump, entire Energy Roof System, and domestic solar water heater)	
Total building UA	400 Btu/ °F. hr.
Heating	
Area of solar heat collection	1,500 sq. ft.
Solar heating	100 percent
Auxiliary heat type	electric heat pump
Auxiliary heat need	has not been used through four years
Cost of auxiliary heat	$0
Cooling	
Area of passive-heat dissipator	1,500 sq. ft.
Auxiliary cooling type	electric heat pump
Auxiliary cooling needed	15–20 percent in 3 yrs.
Cost of auxiliary cooling	$150/yr. (average)
Hot water	
Need	27,375 gallons/yr. @ 120 °F. (incoming 65 °F.)
Auxiliary electricity needed	$14/yr.

ANNUAL ENERGY SUMMARY

Nature's contributions	
Heating	9.3×10^6 Btu
Cooling	31.8×10^6 Btu
Hot water	3.2×10^6 Btu
Total household purchased energy	
Electricity	32.2×10^6 Btu
Energy Roof operation	5.35×10^6 Btu
Auxiliary cooling	8.54×10^6 Btu
Auxiliary hot water	$.79 \times 10^6$ Btu
All other household needs	balance
Total purchased electricity	32.2×10^6 Btu or 21,488 Btu/sq. ft.

Water on the roof for passive cooling as well as heating is a fascinating subject whenever alternative, low-cost space conditioning is discussed. And for good reason: these systems can deliver total home comfort, in both winter and summer, throughout most of the southern and southwestern regions of the United States. But, despite the enthusiasm for roof ponds among passive solar advocates, few have been constructed. This modest Arizona home, in one of the hottest deserts in the world, is one of the few operating roof pond buildings. It has been lived in continuously since 1978, and it was built without any government stimulus or financial support.

Although the flat, 10-acre home site is in a harsh, untouched and rather desolate desert, to its owners, it was ideal. "We picked the site for its unencumbered expanse of view," Frank Fry said. "The mountains dominate the view from the house and are seen from almost every window, ever changing with shadow and subtle color. The desert itself rolls gently and is alive with surprizingly fast growing and changing vegetation and flowers. It is an ideal place to meditate in calm after a hectic day of work."

A Formidable Challenge

The hot desert climate presented a formidable challenge to the young architect who designed the Fry residence. The cooling loads on buildings in this region are typically more than four times the heating loads. Guilford Rand, one of the few designers to have worked on roof pond buildings, has built more than 40 solar buildings of many types throughout the southwest. In 1976, he became aware of a patented roof pond system known as the Energy Roof. This roof provided 100 percent heating and cooling for a 300-square-foot test building at Arizona State University. The Frys visisted the building and subsequently asked Rand to use the system in their new home. When built, the design had a construction cost of less than $45,000, a low figure for a house this size.

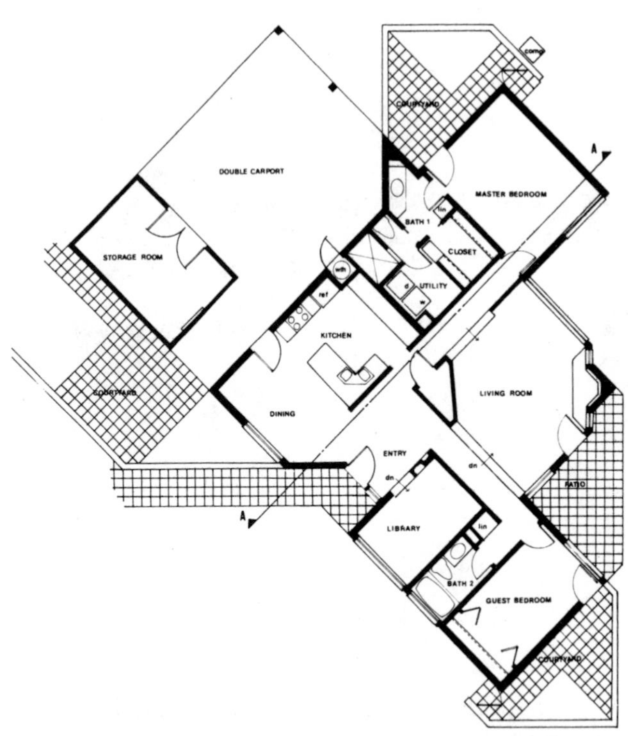

Design of this house encourages outdoor living, with its three courtyards and a patio, all offering views of desert and mountains.

A Contemporary Desert Design

To make the Fry house compatible with a desert setting, a design was developed with soft rounded corners and edges and with walls that extended out to enclose private courtyards. The courtyard walls helped to lengthen the building and draw it back down into the landscape. The flat roof and the light, sand-colored, stucco walls blend easily with the desert colors. There is little hint that this is a solar home.

"In addition," Rand said, "the design was intended to encourage outdoor living during those times when desert conditions are delightful. An open southwestern court for winter enjoyment, and a shaded northeastern patio for the summer were included. The result was a very functional home designed around the living patterns of the owners that harmonized with the harsh desert setting and yet offered protection and comfort for the Frys."

Roof Pond and Design

Another premise was that the most feasible solar system for both heating and cooling was a roof-pond type. In several ways, the roof pond influenced and enhanced the attractive design of this home. For example, in extreme climates such as that found in Arizona, roof ponds can provide 100 percent heating and cooling, but only in one-story structures. On the other hand, there is the inherent flexibility: while such systems require flat roofs, orientation is not important. Thus, the rooms and their openings can be selected to take best advantage of views and shading.

Although the plan of the house may look complex, the opposite is true. The house has a simple, compact, rectangular design. But, this does not limit views and openings in all directions. As is typical of climate-responsive houses in the desert, the windows areas are relatively small. Only 150 square feet—10 percent of the 1,500 square foot floor area—are devoted to windows. And all windows are shaded in midsummer.

The step-down living room is the largest interior space. There is a raised fireplace for cold winter nights. Perhaps the most distinctive aspect of the interior is the fluted steel roof deck, which is exposed as a ceiling throughout the house. The metal decking has been left in the natural, factory-finished metal color, without paint or covering. Since the roof is the heat collector in winter and heat dissipator in summer, the metal ceiling, which serves as a continuous radiator, is essential to the system.

How the Pond Was Built

The pond and steel ceiling/roof are supported by 2 × 6 stud walls, insulated with R-19 fiberglass batts, and built up on an uninsulated, concrete-slab foundation. The 6-inch water pond, which covers the entire roof, lays within a parapet and does not extend beyond the line of the perimeter walls. Porch roofs and overhangs are constructed separately. In effect, the roof of the house is formed into a shallow basin. A liner of reinforced polymeric sheet by *Shelter-Rite* is fitted and installed inside the parapeted area as the basin waterproofing. It is then filled with 6 inches of water. Two layers of 2-inch thick, floatation-grade, extruded polystyrene foam-insulation board are then laminated together and cut to fit the shape of the pond area. This floating platform has a 2-inch clearance all around the perimeter.

A water dispersement system of small, perforated plastic tubes is then spread out horizontally on top of the floating insulation and connected to a small pump. Another reinforced, grey, polymeric sheet is stretched over the entire pond area and fastened and sealed around the edges. This encloses the pond system completely and also sheds rain water.

How the Pond Roof Works

Winter Heating Cycle. Architect Rand explains the operation of the roof this way: During daylight winter hours, water is pumped from the roof pond up between the polystyrene foam insulation and the gray polymeric top cover. The water spreads out in a thin layer beneath the polymeric sheet and the insulation surface, and is then heated by the sun. By the time the water reaches the edge of the insulation, it is warm and trickles back into the pond. The solar warmth is stored in the pond until the indoor temperature begins to cool, at which time the warm water radiates its heat via the steel ceiling into the space. The steel decking acts as a radiant ceiling providing the most efficient, quiet warmth possible.

Summer Cooling Cycle. During the summer, the pond of water must be cooled so that it may act as a heat absorber inside the house. The process of cooling the water in the pond is the same as for heating it, except that it is done at night. The cool water in the pond absorbs heat generated in the house. At night, the pond water is pumped between the floating insulation and the top cover. The house-warmed water radiates its heat into the night sky and is cooled to within 10 degrees Fahrenheit of the wet bulb temperature. Usually, this is cool enough to maintain comfortable interior spaces.

Strictly speaking, this passive solar heating and cooling system might be classed as a hybrid because of its use of a ¼ horsepower pump. However, the expenditure of 535,000 Btu of electric energy for pumping produces 31.8×10^6 Btu of cooling or over 50 times the energy purchased. While not a free ride, this coefficient of performance is an enviable achievement.

Staying Cool in a Hot Climate

In the hot, arid southwest deserts and highlands, this roof system can be an ideal way to minimize residential energy consumption. During the first three years of operation, the Energy Roof has provided 100 percent of the space heating and 80 percent of the cooling for the Fry residence. It is cost competitive with conventionally built and air conditioned structures. Snowfall poses the only real limitation. In more northerly climates, heavy snowfalls might shut down the system temporarily, until the snow melted. But, this is not a concern in snowless Yuma, Arizona. Here a pleasant and charming rural house gives almost no hint of its pace-setting energy performance.

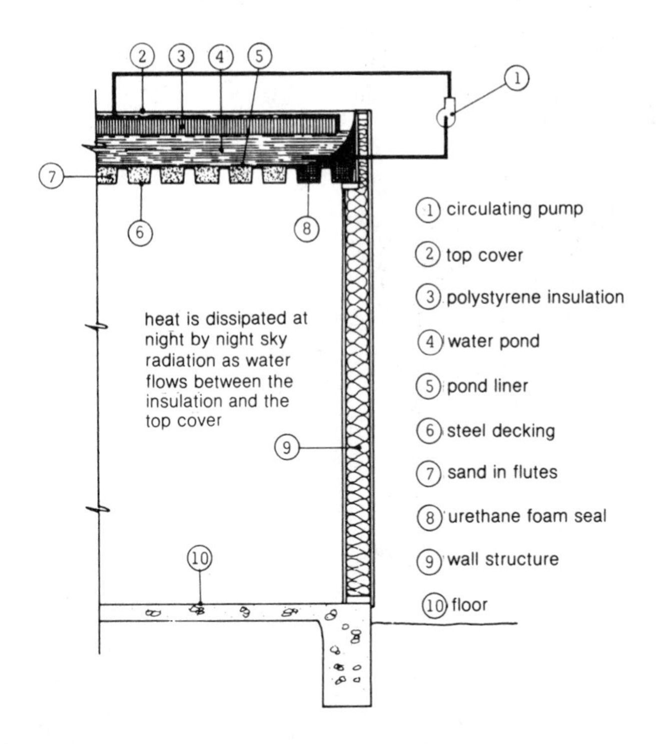

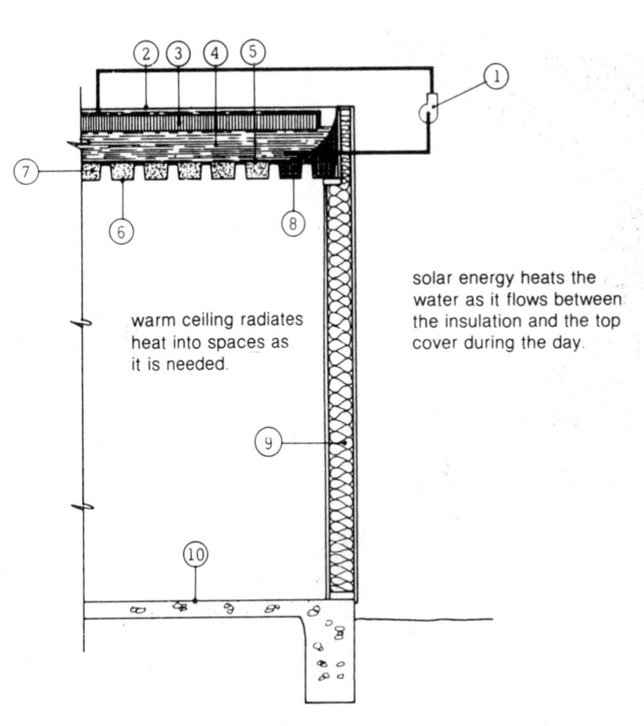

Roof pond helps to cool house in summer (top), warm it in winter.

Passive Solar Homes For A Sacramento Subdivision

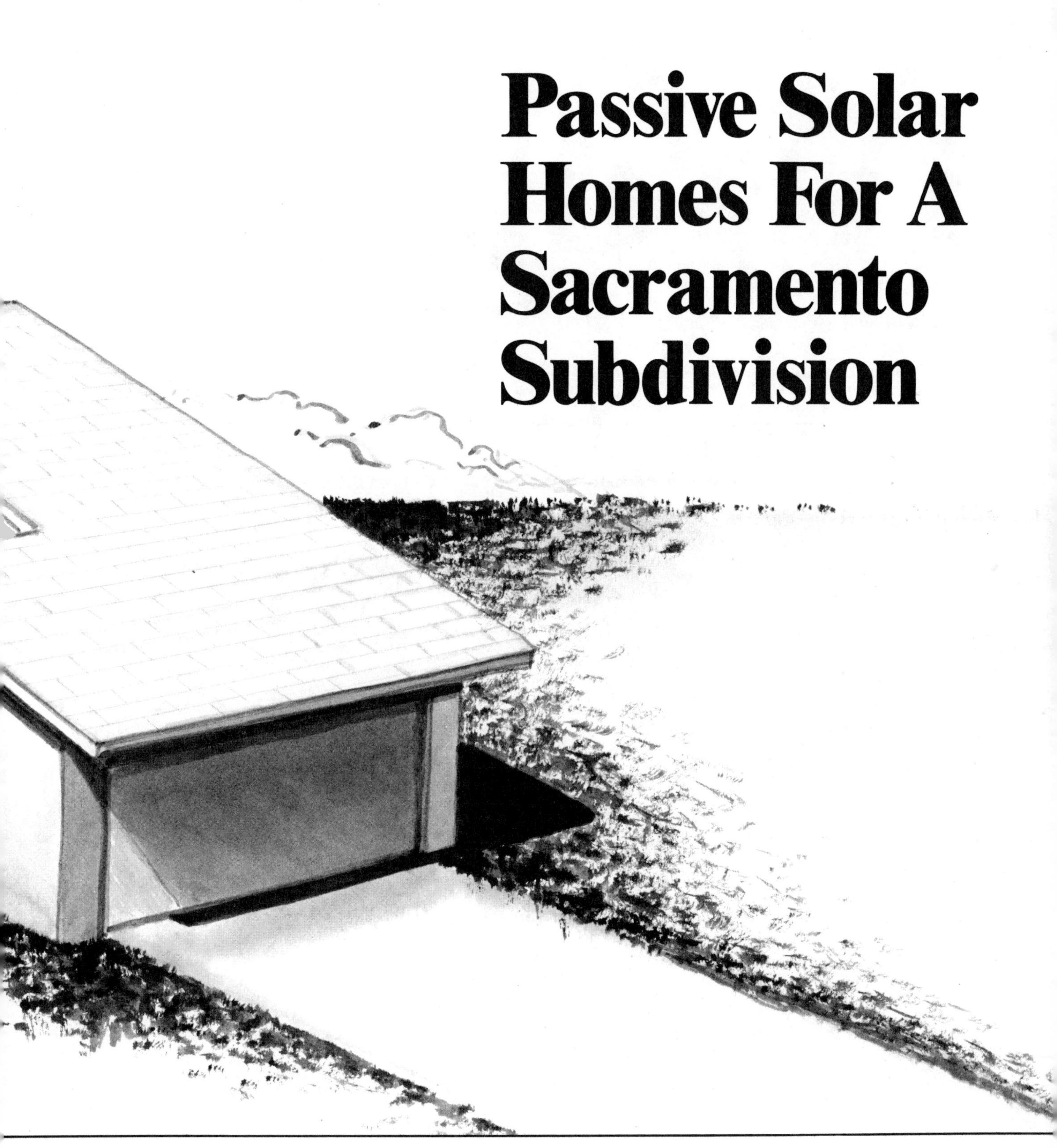

PROJECT FOR

Larchmont Passive Solar Subdivision
Sacramento, California

DESIGNER

Charles Eley Associates
342 Green Street
San Francisco, California 94133

BUILDER

M. J. Brock & Sons
3350 Watt Avenue
Sacramento, California 95821

STATUS

Unbuilt

Estimated Building Energy Performance
(1,540-Square Foot Larchmont Tract House)

BUILDING DATA

Heating degree days	2,843 DD/yr.
Cooling degree days	1,159 CDD/yr.
Floor area	1,540 sq. ft.
Estimated building sales price	$95,000
Cost of special energy features	$3,000
Total building UA	603 Btu/°F. hr.
Heating	
Area of solar-heat collection	177 sq. ft.
Solar heat	45 percent
Auxiliary heating	heat pump (COP 2.5) and forced-air distribution
Cost of auxiliary heat	5.7¢/sq. ft./yr.
Cooling	
Natural convection	83 percent of total load
Auxiliary cooling	heat pump (COP 2.5)
Cost of auxiliary cooling	2¢/sq. ft./yr.
Hot water	
Need	29,200 gallons/yr. @ 110°F. (incoming 65°F.)
Cost of auxiliary gas	$9/yr.

ANNUAL ENERGY SUMMARY

Nature's contributions	
Heating	6.0×10^6 Btu
Cooling	12.6×10^6 Btu
Hot water	7.7×10^6 Btu
Auxiliary purchased	
Electricity	27.0×10^6 Btu or 17,500 Btu/sq. ft.

Those who can afford custom-built houses sometimes scorn the ubiquitous tract house as anonymous, poorly designed and badly built. And it is true that to produce economical developments there must be repetition of identical houses, though with some variety of house size and layout to satisfy different family needs.

Yet both functional design and sound construction can benefit from the production economics of serial house construction. The economy of tract housing is obvious to anyone who has tried to buy land and hire all the services to build a custom house. The tract house represents a bargain at every level of expenditure. One can buy more space and more quality for the same money in a tract house.

The passive solar houses planned for the Larchmont subdivision in Sacramento are good examples of balancing production economy and human interest. The sun-flooded entry hall and the lines of pot sills on the exterior are features that would be admired apart from their energy purposes. The orderly arrangement of land planning, structural bearing lines and elevations are part of the architect's design discipline. While they are important, the passive heating and cooling techniques do not dominate the design. Within the common window sizes, structural systems and lot setbacks, there is a variety that will make Larchmont a pleasant subdivision that integrates solar energy into a larger housing goal.

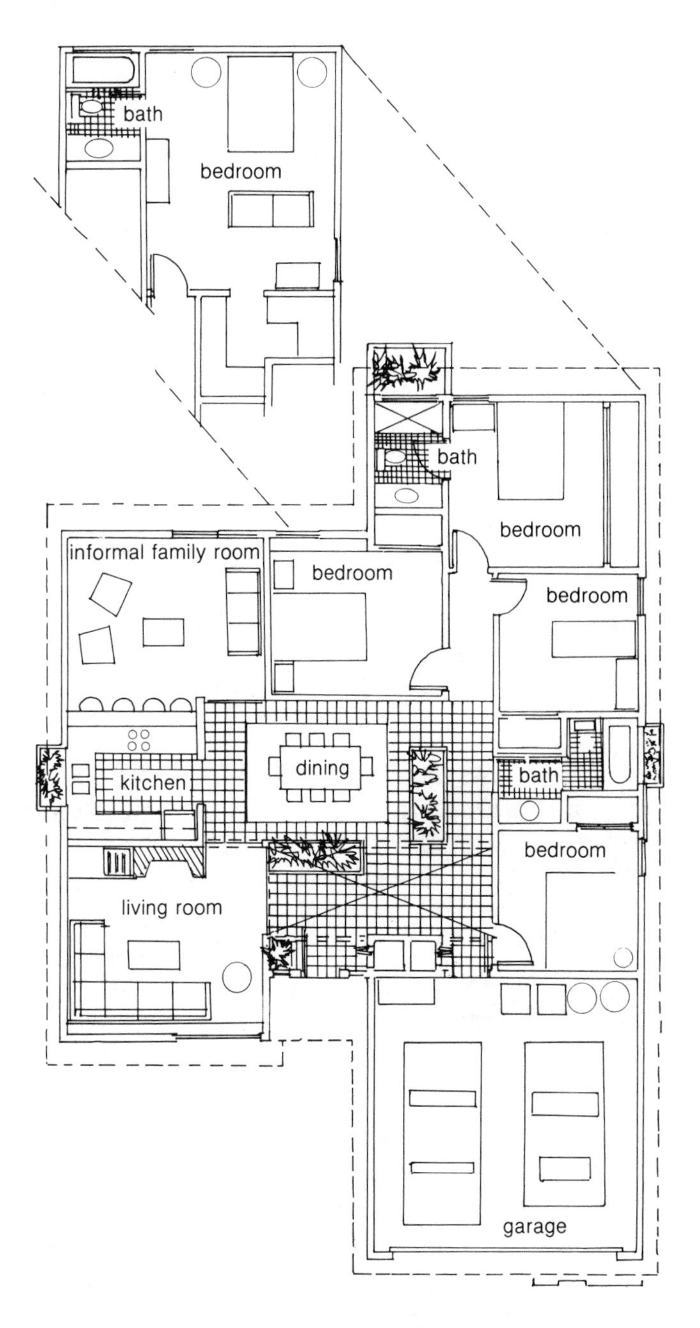

This tract house was built to rigid criteria. Drawing at top of floor plan offers an alternate, a large bedroom instead of two smaller ones.

The Solar Challenge

From a solar viewpoint, tract developments are a special challenge. All house sites are not on the south side of the street. In many subdivision layouts, in fact, it is necessary to have house designs that can face at least the four cardinal directions and sometimes other orientations as well. When the solar requirements are combined with a desired variety of house sizes, the possible variations within a subdivision may seem infinite. So the design challenge is to develop solar houses for multiple orientations and to design several house plans with a minimum number of prototypes.

The four house designs for Larchmont are different but related, though each has its own character. Their character is more developed than typical

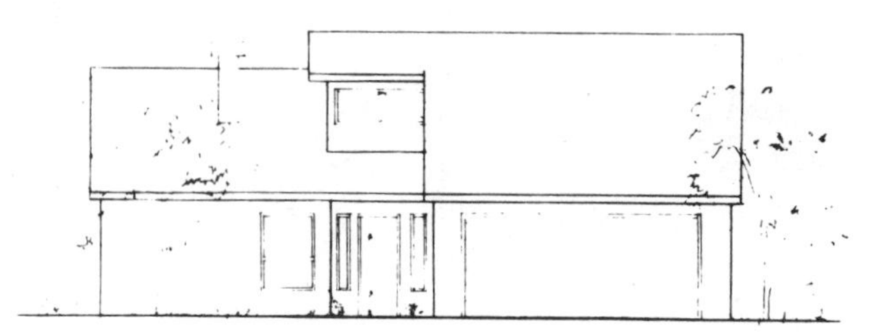
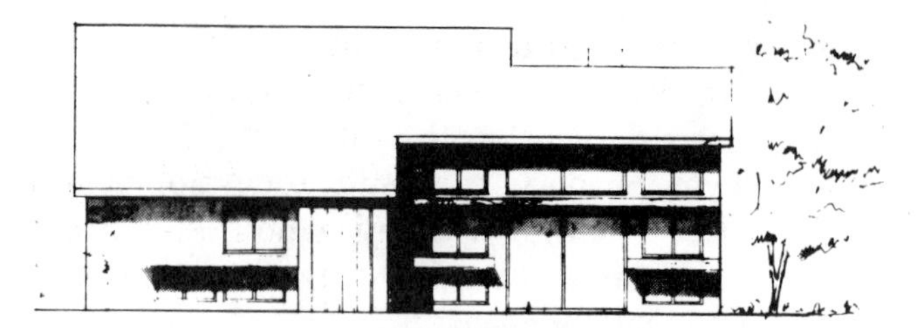

North (left) and south elevations.

tract designs in most other locations, partly because they share a reliance on passive heating and cooling.

Matching the Needs of Builder and Climate

M. J. Brock & Sons is a large and successful builder of tract housing. For the Larchmont design, they chose the architectural office of Charles Eley, a San Francisco firm that has specialized in providing design services for tract builders. As a client, the Brock organization was interested not only in a design for the proposed Larchmont subdivision but also in including passive solar features in their entire housing line—without impairing cost and marketability. They established several mandatory criteria at the outset of the design process:

- Each home must be conventional in appearance and have a two-car garage.
- Each floor plan must function on both the north and south sides of the street.
- Conventional wood trusses must be used for ease of construction and low cost.
- Although the floor is slab-on-grade, carpet must be used in all living areas. Hard floor surfaces are permitted only in the kitchen, entry, and dining areas and corridors.

The lots for these homes are located on east-west streets and are relatively small—52 feet wide by 100 feet deep. Five-foot side yards and 20-foot front and rear yards are required, leaving buildable envelopes of 42 by 60 feet.

The Sacramento winters are fairly mild, with 2,843 heating degree days. Summer days typically are in the 90s, although evenings generally cool off to below 60° F. Summer breezes are from the south. This climate calls for designs that provide effective heating together with cooling.

A Versatile and Efficient Layout

The Larchmont houses range in size from 1,020 to 1,540 square feet, and have two to four bedrooms. Since each of the house types could face either north or south, a total of eight designs could be seen along any street.

The interior layout reflects a California lifestyle with the varying character of different spaces. The kitchen is a large and efficient U-shape that opens across the breakfast counter to an informal family room. But the dining room is formally arranged in the center of the house, partially defined by built-in planters.

The living room is away from the main circulation of the house and has a large masonry fireplace for use on cold winter evenings.

Heating and Cooling

During the heating season, the house uses a combination of direct-gain and isolated-gain methods. When the design is located on the north lot, a tall entry area is used to obtain the needed south-facing glass. This tall space also provides a stack effect for better natural ventilation. The dining room/ kitchen/family room area is located on the opposite side of the lot from the garage so that the sun

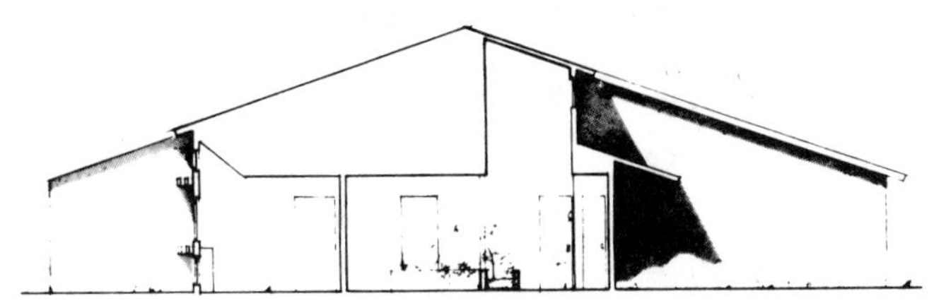

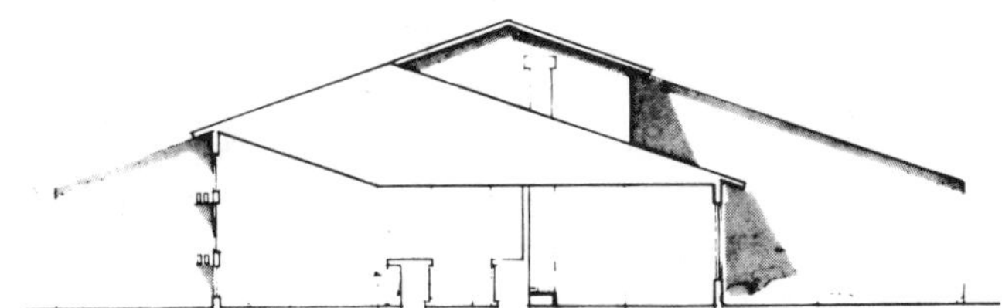

Two east-west sectional drawings.

can penetrate through either the front or rear yard. Hard surfaces and planters provide thermal mass throughout this area. A counter-height Trombe wall is incorporated in the dining area (north lot only) or the family room and bedroom areas (south lot only), collecting heat by day and releasing it at night. The walls are shaded by small pot shelves. Thermal mass for the direct-gain system is also provided by an interior masonry fireplace enclosure and by the slab itself where it is exposed in the dining and kitchen areas.

During the cooling season, night ventilation cools the house and lowers the temperature of the interior thermal mass. Air is moved by a combination of wind and stack effects. During the day, the house should be sealed and the drapes closed. Predictable high daytime temperatures and low night temperatures can thus be controlled if summer shading prevents large daytime heat gains.

Active solar water heaters are used with collectors located on the garage roof. By locating the insulated hot water storage tank and the laundry in the garage, their heat and moisture are kept outside the living spaces.

Performance calculations have been made using the California Passive (CALPAS) computer program, assuming conventional levels of insulation—R-11 wall, R-19 ceiling, and double-glazed windows. When the homes are ultimately built, however, the ceiling insulation will be increased to R-30.

A Passive Solar Subdivision

Beyond the interesting architectural character of individual houses and their specific thermal expectations, are the demands of the neighborhood layout. In general, the land planning of housing subdivisions in California involves concepts that conserve resources as well as maximize the usability of outdoor space. Fundamentally, this is done by planning for higher density, which increases the yield of house lots and thus reduces the length of street paving, sewer and water lines and other improvements that must be paid for by each house owner. When houses are to be solar, the orientation of every house and its future access to the sun are critical, and become an added requirement in subdivision planning. In Larchmont, all these needs are met.

In this subdivision, the shape of each house is arranged to define exterior spaces. In the front, projecting garages define the entry garden; in the rear, the walls and rooms help articulate the backyard. Thus, each house has a clearly established identity along the street that focuses on the front door, and a well-defined private yard in the back. Although the house lots are small, they have an implied use plan for each homeowner to develop his own individual landscaping scheme.

The California subdivision pattern of fully landscaped yards and designed outdoor uses has been widely emulated elsewhere. It is an approach that maximizes the usefulness as well as the appearance of expensive land and small house lots. Together with the stringent orientation requirements of solar houses, the Larchmont design demonstrates how a passive solar subdivision can provide a lively and economical neighborhood within the constraints of production construction and solar geometry.

Passive Solar Modular Homes

Estimated Building Energy Performance
TVA Solar Modular Homes–Two Bedroom/Solar Crawl Space

BUILDING DATA

Typical performance calculated for Memphis, Tennessee

Heating degree days	3,227 DD/yr.
Cooling degree days	2,029 CDD/yr.
Total ground area	912 sq. ft.
Net conditioned area	806 sq. ft.
Building cost	$28,000 to $35,000
Cost of special energy features	$3,000
Total building UA (day)	434 Btu/°F. hr.
Total building UA (night)	236 Btu/°F. hr.

Heating

Net area of solar-heat-collection glazings	127 sq. ft. direct gain and 64 sq. ft. crawl space
Solar heat	60 percent annually
Auxiliary heat	electric heat pump (COP 2.7)
Auxiliary heat needed	4,219 Btu/sq. ft./yr.
Cost of auxiliary heat	1.4¢/sq. ft./yr.

(Optional wood-burning heater not included)

Cooling

Ventilated attic space
Natural cross-ventilation
Auxiliary cooling by electric heat pump

Cost of auxiliary cooling needed	none

Whole house fan

Hot water

Need	25,500 gallons @ 120°F. (incoming 60°F.)
Auxiliary gas cost	$46/yr.

ANNUAL ENERGY SUMMARY

Nature's contributions

Heating	17.5 × 10⁶ Btu
Cooling	not calculated
Hot water	70 percent or 10.7 × 10⁶ Btu

Auxiliary purchased

Electricity and gas	25.3 × 10⁶ Btu or 27,700 Btu/sq. ft.

Estimated Building Energy Performance
TVA Solar Modular Homes–Four Bedroom/Solar Great Room

BUILDING DATA

Typical annual performance calculated for Memphis, Tennessee

Heating degree days	3,227 DD/yr.
Cooling degree days	2,029 CDD/yr.
Net conditioned area	1,406 sq. ft.
Building cost	$40,000 to $50,000
Cost of special energy features	$3,000
Total building UA	649 Btu/°F. hr.
Total building UA (night)	334 Btu/°F. hr.

Heating

Area of solar-heat-collection glazings	281 sq. ft. direct gain and 70 sq. ft. greenhouse
Solar heat	78 percent annually
Auxiliary heat	electric heat pump (COP 2.7)
Auxiliary heat needed	2,414 Btu/sq. ft./yr.
Cost of auxiliary heat needed	.8¢/sq. ft./yr.

(Optional wood-burning heater not included)

Cooling

Natural cross-ventilation
Auxiliary cooling by electric heat pump

Auxiliary cooling needed	not calculated
Cost of cooling	not calculated

Hot water

Need	43,800 gallons @ 120°F. (incoming 60°F.)
Auxiliary gas cost	$61/yr.

ANNUAL ENERGY SUMMARY

Nature's contributions

Heating	42.5 × 10⁶ Btu
Cooling	not calculated
Hot water	1.83 × 10⁶ Btu

Auxiliary purchased

Electricity and gas	32.6 × 10⁶ Btu or 23,200 Btu/sq. ft.

PROJECT FOR

Tennessee Valley Authority (TVA)
Solar Applications Branch
Chattanooga, Tennessee

DESIGNER

Michael Sizemore
Thomas Sayre, Project Manager
Sizemore/Floyd
Architects and Energy Planners
1 Piedmont Center #200
Atlanta, Georgia 30305

BUILDER

Guerdon Industries
Box 35290
Louisville, Kentucky 40232

STATUS

Unbuilt

The Tennessee Valley Authority is one of the world's largest electric utilities. As part of a comprehensive power-management plan, the TVA commissioned major mobile home and modular home manufacturers in the region to create a series of economical solar-housing designs suitable for the TVA region. Guerdon Industries, a successful modular home manufacturer, collaborated with Sizemore/Floyd, architects and energy planners, to design two different passive solar, modular homes.

These modular houses are derived from a systems approach of interchangeable parts and components. Thus, they can offer a surprising variety of custom-assembled features. That they are pre-engineered and factory assembled means that these homes can offer a broader range of choices and a potentially more dependable housing product than many custom builders can offer and the price should be much more economical.

Aesthetically, these designs are no one's dream cottage. They are "Plain Jane" solutions to housing needs as preferred by their potential purchasers. Maximum advantage, for instance, has been taken of the standard 12-foot and 14-foot wide modules. The dimensions of all spaces have been designed for maximum useability. But choices of exterior finish and material, landscaping and interior furnishing are personal preferences that are not controlled by the design decisions of the system. Especially in the gentle, moist climates of the southeastern states, vegetation, more than any other aesthetic consideration, will soften and personalize each of these modular homes.

Solar Housing at a Modest Cost

Because builders who supply the housing market find higher-priced construction more profitable, middle-income families in the TVA area and in neighboring southeastern states often find their housing needs ignored. The design preferences of this market are low cost and low operational energy costs within a traditional imagery. This conservative market welcomes the use of solar as a means of maintaining low utility bills.

The housing units shown here were priced in the $30,000 to $50,000 range in 1980, and designed for acceptance by Veterans Administration and Federal Housing Administration financing. Industrial plants in the region can produce thousands of units per year before reaching production capacity, and each house can be assembled and be fully operational in three weeks. These modular units move passive solar housing out of the exclusive domain of the rich and self-reliant and into the reach of the middle class.

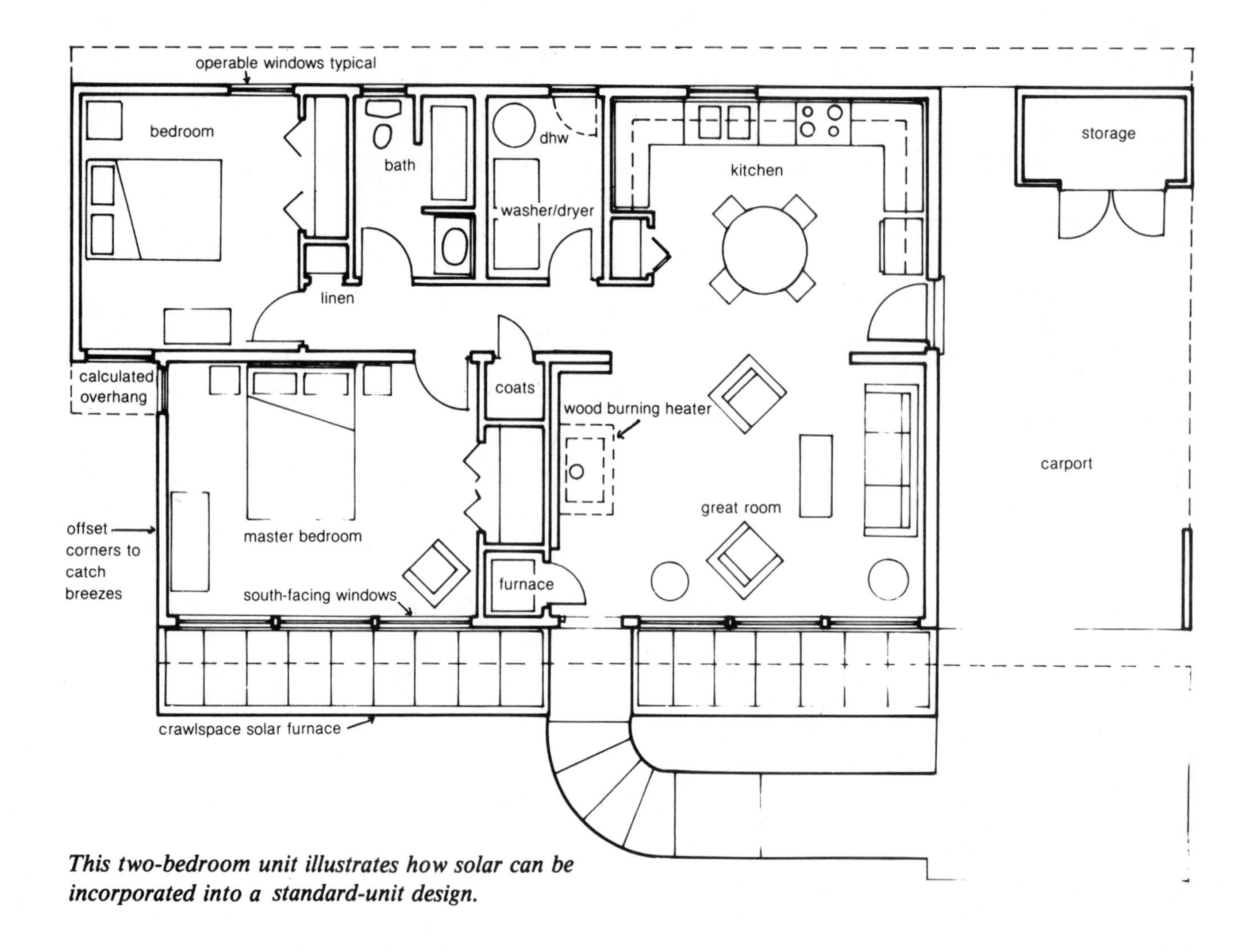

This two-bedroom unit illustrates how solar can be incorporated into a standard-unit design.

The Two-Bedroom Design

To satisfy this growing market, the designers have created two affordable houses that combine state-of-the-art solar technology with premanufactured construction. Least expensive is the two-bedroom model, which essentially incorporates solar into a standard-unit design.

In creating a solar two-bedroom model, the major design effort was to develop a crawlspace heating system. The solar crawlspace allows the building to keep a traditional appearance and still provide significant annual solar heating. Heat stored in fixed water-filled thermal storage bags (similar to water beds) can rise passively through ducting to warm the living space. When no heat is being used, modest losses from thermal storage keep the house dry. The solar heat that warms the water bags passes through angled diffuse glazing at the front of the house; night insulation reduces heat losses.

Besides its traditional look, the "road appeal" or visual marketing attractiveness of the two-bedroom unit is increased by offset corners that act as wind scoops oriented to the prevailing breezes. Multiple-entry orientations are available without a change in the building site orientation, so several types of sites can accommodate this solar design.

The Top-of-the-Line Model

The four-bedroom design is the top of the line. It combines site-built construction with pre-manufactured economy: all doors, electrical systems, mechanical systems and plumbing are installed in the factory, thus keeping the benefits of high quality and low cost available in modular housing. A large, four-bedroom unit can be shipped on a single truck, drastically cutting transportation costs.

Unlike the two-bedroom house, the four-bedroom unit has a site-built solar great room that incorporates thermal storage mass and additional solar glazing. Windows from an overhead clerestory allow deep light penetration and free the central space from the transportation height limits placed on modular units. It also gives a brighter, more open space as a central area than does the customary low-ceiling modular design. The clerestory, sunspace and offset corners add to the "road appeal" and the distinct solar image of the four-bedroom house.

Solving the Thermal Storage Problem

Because of its need to be transported, modular housing is characterized by lightweight construction and lack of thermal mass. While the moderate amount of solar glazing can be factory installed, the thermal mass needed to moderate interior temperature fluctuations must be installed on site. For this reason, all thermal mass is composed of locally-available materials—water bags in the two-bedroom unit and brick pavers over concrete slab in the four-bedroom unit (the manufacturer precluded the use of eutectic salts and other less traditional thermal-storage materials.) The thermal mass is installed after the houses have been placed on their permanent sites. (For complete specifications for this home, see Appendix B.)

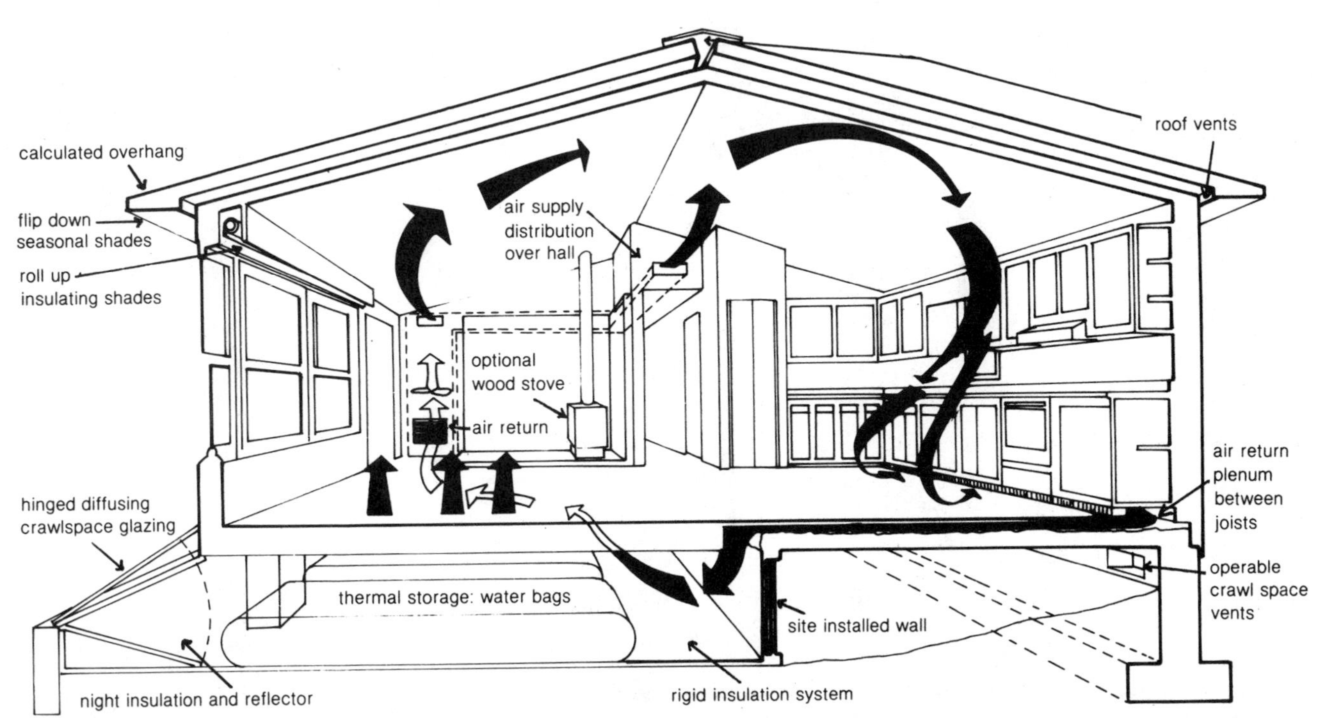

The pattern of the movement of heat through the crawlspace heating system in the two-bedroom home is shown. Note water-filled thermal storage bags under floor.

Isolated Thermal Storage

Isolated thermal storage, shown in the two bedroom design, has the advantage of maximizing the opportunities for control. It can be used in various ways. The air-distribution system is connected with the thermal storage crawlspace through a mixing box. When auxiliary heating or cooling is engaged, the mixing box only draws air from the rooms. Night thermostat setbacks can be used effectively because the mass is not in direct contact with the spaces being heated and cooled.

Under moderate heating loads, a damper is closed and air is drawn from the warm crawlspace in a "fan only" mode. At night, the R-15 interior insulating shades and the insulating crawlspace shutters are closed and stored heat can be moved into the house.

Windows in the living spaces are sized to satisfy average day time heating needs during a mid-winter day. Any excess heat is circulated past the storage mass by the fan.

In the cooling season that includes parts of spring and fall as well as summer, the offset house corners aid in cross-ventilation from prevailing north or south winds. Further cooling is provided by a whole house fan and by oversized roof vents to reduce roof heat build up. Light-colored roof surfaces and reflective exterior surfaces on all operable insulating shutters drastically reduce heat gains through the exterior skin. These systems, plus the forced circulation of night air past the shaded storage mass, make auxiliary cooling unnecessary.

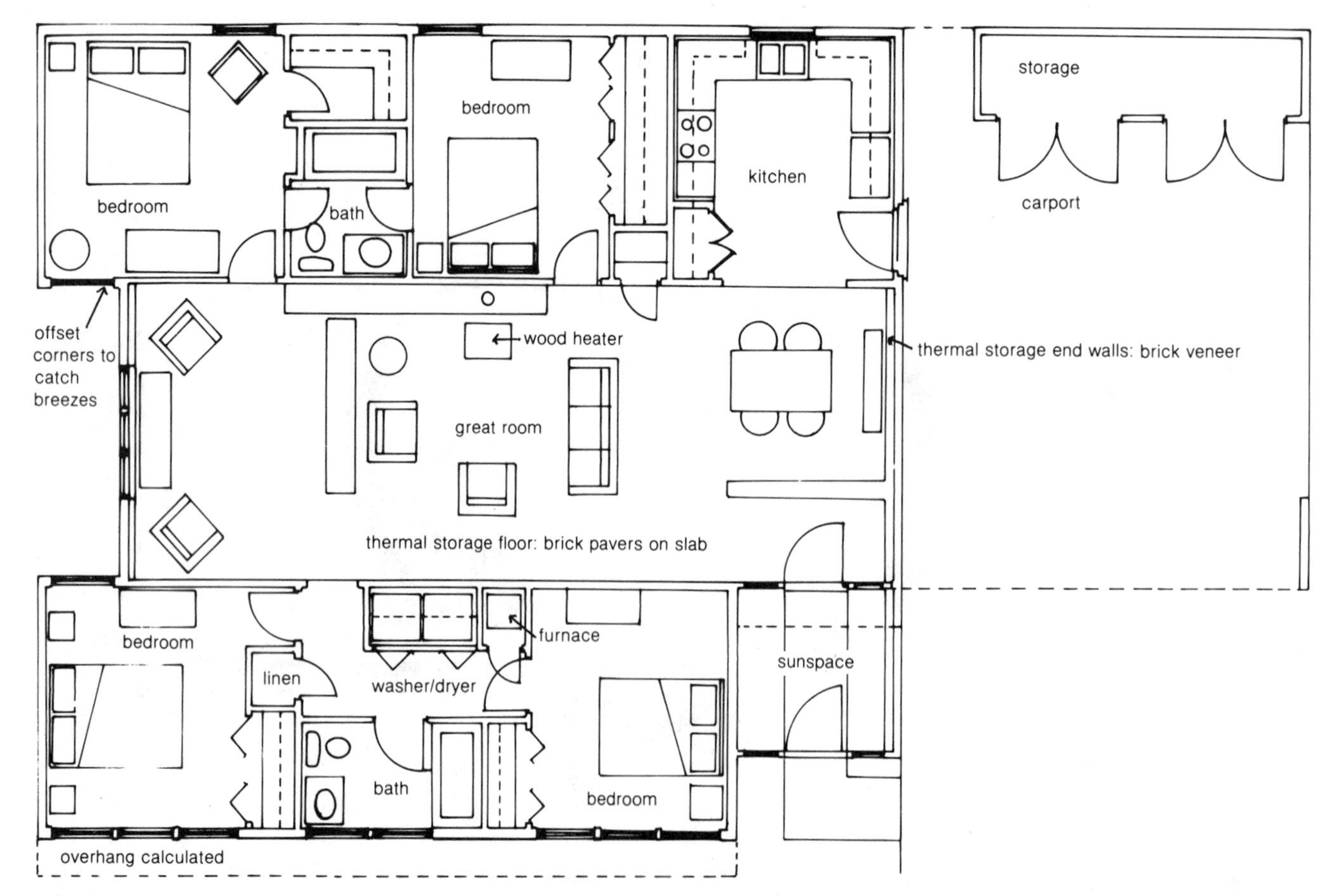

The four-bedroom design features a clerestory to allow light into the central living area.

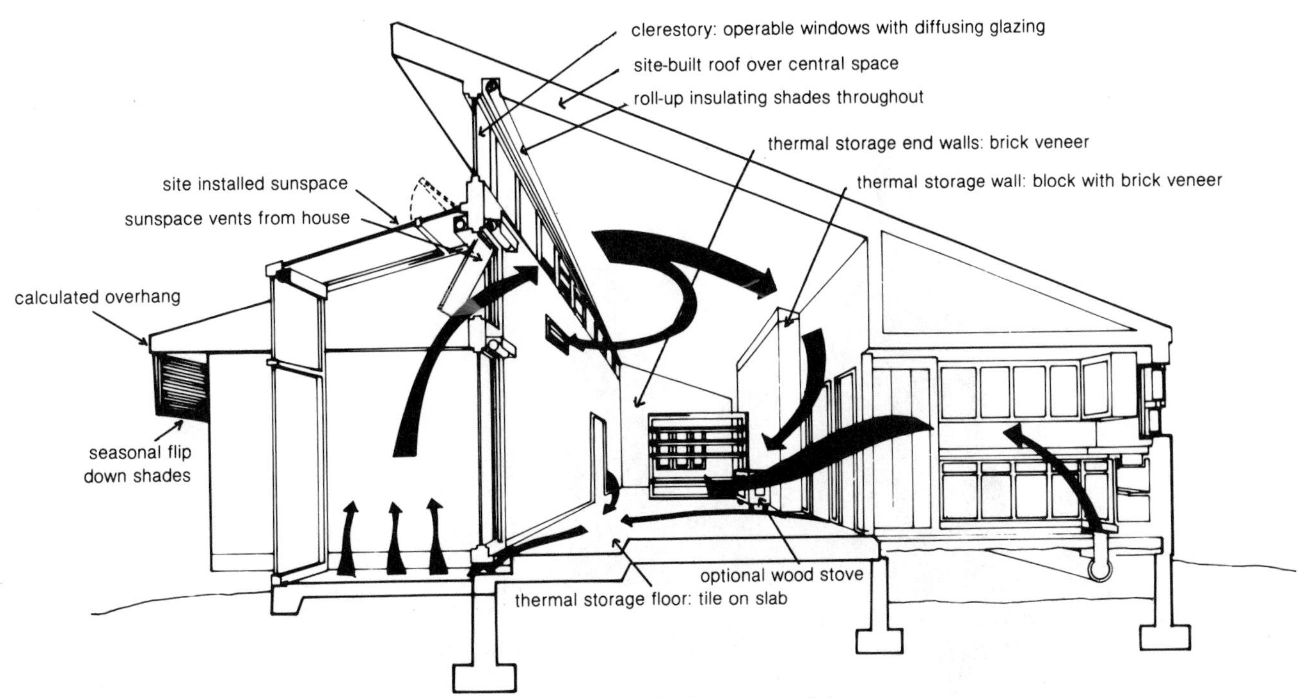

This is the flow of heat in the heating mode in the four-bedroom model.

In typical houses in the TVA southern regions, air conditioning cooling costs rival heating costs, but few homes are designed to solve the dual problem of heating and cooling. In addition to the visible design strategies, these homes also use air to air heat exchangers to maintain interior air quality. Exchanges are needed due to the low infiltration of these tightly constructed houses.

Stepping Up to Central Thermal Mass

The four-bedroom house has all of its thermal mass in the central, site-built section. The masonry storage mass—a slab floor with brick surface and brick end walls—is enclosed between two tight, well-insulated modular units. The tall central space has diffusing clerestory glazing that distributes much of the solar heat directly to the thermal mass. The central space is also tall enough for warm air to stratify. Warm air is ducted back to the living space through a grille located high in the clerestory.

In the cooling mode, operable clerestory windows exhaust much of the hot air. Circulating fans, which help to destratify air during the heating season, can be run at a higher speed to circulate air and cut cooling requirements. As in the two-bedroom design, reflective R-15 night insulation drastically cuts heat transfers and reflects much of the summer solar heat gain.

Two Models Built

Two models of the Guerdon Industries houses have been built in Chattanooga—one at the TVA Solar Test Center, the other at a TVA training center. Because of a decline in the economy, however, this ambitious housing program based on production economies has not moved ahead as planned. When it does, this design concept will have a measurement of its real worth—the quantity of houses that are actually sold and used. Until then, the potential solar contribution of 60 percent that these designs promise will remain only a potential, yet one that could have a tremendous impact on the energy use of the Tennessee Valley.

Low-Cost, Panelized TVA Houses

PROJECT FOR

Tennessee Valley Authority
Solar Applications Branch
Chattanooga, Tennessee

BUILDER

Panelfab International, Inc.
P.O. Box 2777, AMF
Miami, Florida 33159

DESIGNER

Michael Sizemore
Sizemore/Floyd Architects
and Energy Planners
1 Piedmont Center
Atlanta, Georgia 30305

STATUS

Unbuilt

Estimated Building Energy Performance

TVA Panelized Solar Home–Two bedrooms with sunspace (three bedroom with solar great room similar)

BUILDING DATA

Typical annual performance calculated for Memphis, Tennessee

Heating degree days	3,227 DD/yr.
Cooling degree days	2,029 CDD/yr.
Net conditioned area	764 sq. ft.
Building cost (with carport and solar domestic hot water)	$35,000
Cost of special energy features	$3,800
Total building UA	348 Btu/°F. hr.
Total building UA (night)	221 Btu/°F. hr.
Heating	
Area of solar-heat collection glazings	144 sq. ft.
Solar heat	50-70 percent (with or without sunspace)
Auxiliary heat type	electric-resistance furnace
Auxiliary heat needed	10,503 Btu/sq. ft./yr.
Cost of auxiliary heat	9¢/sq. ft./yr.

(optional sunspace, wood-burning heater not included)

Cooling

Natural cross ventilation
No auxiliary cooling planned

Hot water

Need	32,850 gallons/yr. @ 120°F. (incoming at 60°F.)
Auxiliary electricity cost	$29/yr.

ANNUAL ENERGY SUMMARY

Nature's contributions	
Heating	7.4×10^6 Btu
Cooling	no calculation
Hot water	9.1×10^6 Btu
Auxiliary purchased	
Electricity	11.6×10^6 Btu or 15,194 Btu/sq. ft.

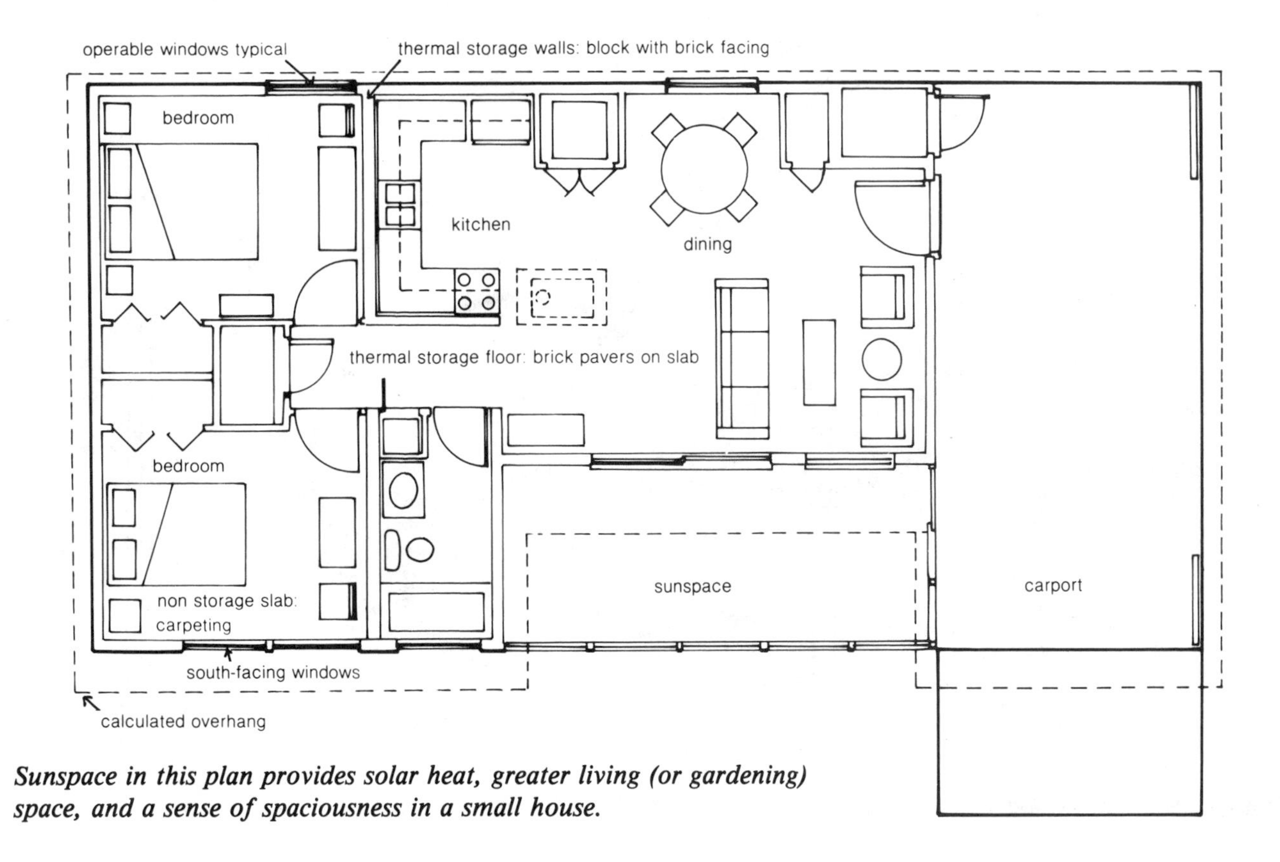

Sunspace in this plan provides solar heat, greater living (or gardening) space, and a sense of spaciousness in a small house.

In 1980, the Tennessee Valley Authority estimated that there was a shortfall of 42,000 new housing units per year within the areas served by the TVA. Housing shortages were even greater in the region immediately surrounding the TVA service area. Population groups with the greatest need were the nonwhite and the elderly. With the average price of a new home above $50,000, new housing was beyond the reach of most individuals and families living in substandard units. Moreover, recent prices for new homes had been increasing at twice the rate of family incomes, and projections were for this trend to continue.

Thus, TVA provided funding to several major housing manufacturers to develop energy-efficient, low-cost housing suitable to the needs of the region. Their goal was to combine state-of-the-art passive-solar design and energy conservation with factory-fabricated, modular building systems to produce high-quality housing units priced between $25,000 and $35,000. TVA proposed an initial demonstration project of 30 units, to be constructed in several different areas of the Tennessee Valley starting in October, 1980. If successful, TVA envisioned building thousands of these energy-efficient units in the region during the 1980s. They would have an important effect on regional energy and electrical use and provide high-quality housing for those for whom conventional standard housing is out of reach.

Despite the creation of a successful design for these units by Sizemore/Floyd, the commercial demonstration units were never built. High capital costs and problems encountered by the panel fabricator in developing a field organization to build on-site thermal mass were the two key stumbling blocks. However, the design provides an excellent model for low-cost solar housing.

Designing Efficient Living Space

To save building and energy costs, the designer planned minimum-size spaces for all housing units

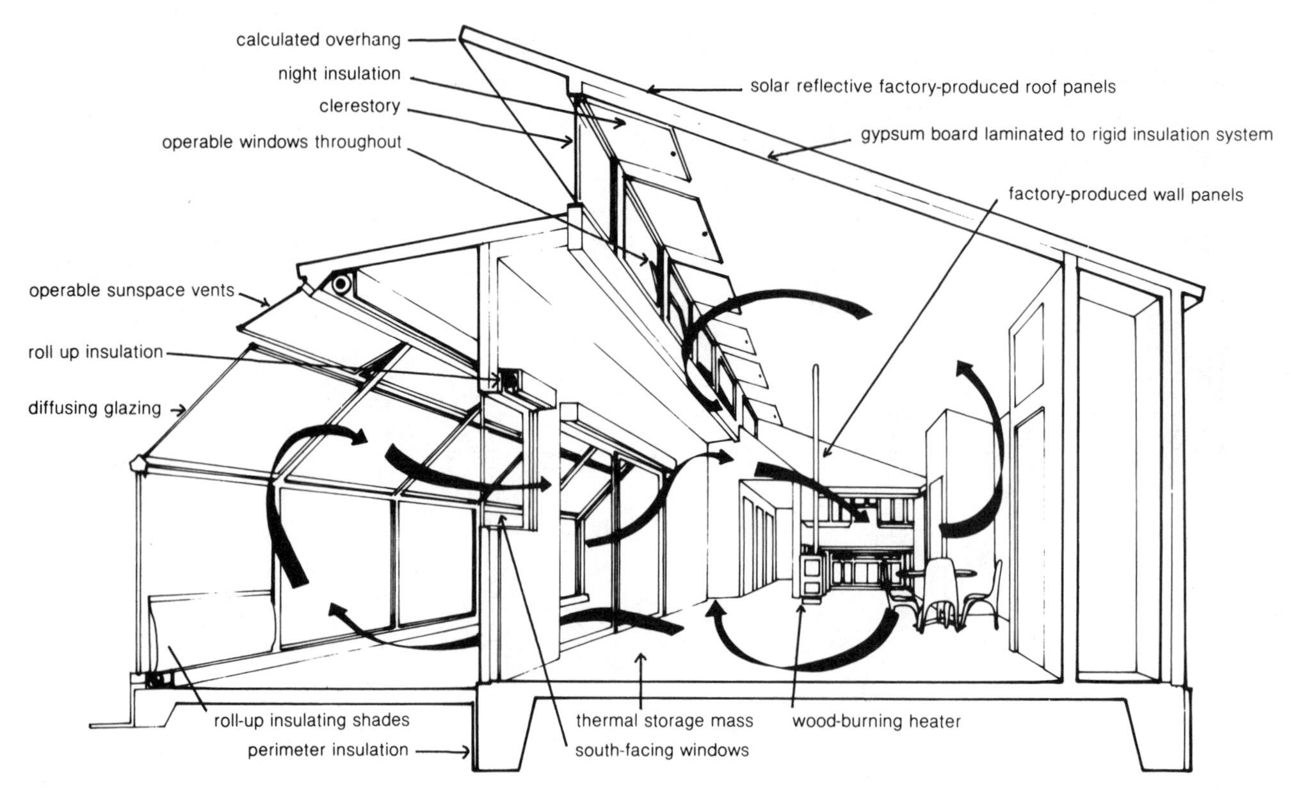

Flow of heat from sunspace is depicted. Note the clerestory at upper left of living area.

—as little as 764 square feet. Two models were chosen—two bedrooms with a sunspace and three bedrooms with a solar great room. Both are organized to be highly efficient in floor space and to make maximum use of a common family and social room. This central space has a high ceiling with clerestory and large expanses of south-facing glass. The floor plan allows south light and some thermal storage mass in every room of the house. The open plan for all units helps to distribute heat and to keep the entire house comfortable.

Creating a House Skin Panel

The housing units are designed to have lightweight, metal-faced, stressed skin panels with honeycomb cores. The construction method used to fabricate these panels was invented by the aircraft industry. The prefabricated panels are assembled on site, using metal extrusions and spline connectors. There is no structural framing. Interior walls, floor slab, plumbing and wiring are also installed on the site. By having the same building cross-section on all models, the panels may be factory produced, thus creating a "family" of housing units that use identical panels. The houses can accommodate larger families or be made more appealing to the market by adding optional features—carports, attached sunspaces and wood heaters.

On the inside walls, electrical wiring is surface mounted. The panels are then covered with insulation and gypsum wallboard. Combined with the exterior connecting panels, this method provides two essentially airtight barriers to prevent air infiltration and movement of water vapor. The closed-cell insulation gives further thermal integrity. By using all electric appliances, air-quality problems that might arise with such an airtight house are minimized.

The external finish on all wall and roof panels is a highly reflective, low-maintenance enamel paint that reduces heat gain during the summer months.

A Direct-Gain, Passive Heating Strategy

To reduce complexity, the designer chose direct gain as the most appropriate passive solar heating strategy for the southeast. The factory-produced building skin and site-installed insulation form a lightweight envelope around walls and floors, which contain the needed thermal storage mass. This means that heat storage is in the center of the house where losses are minimal and the exposed surface is greatest. The centrally located thermal mass also stores heat generated by internal sources, such as ovens and wood heaters. The masonry storage walls and floors are finished with brick pavers or ceramic tile. To further save costs and still permit good energy performance, single-glazed windows with movable night insulation (R-4) are used instead of double glazing.

Solar Domestic Hot Water Included

A thermosyphon system enclosed within the building structure provides solar domestic hot water and significantly reduces the cost and complexity of the hot water system. An electrical-resistance furnace gives auxiliary space heating. Return air ducts to the furnace use manually selected high or low return grills, cutting down on air stratification in the winter and increasing stratification in summer for maximum comfort.

Saving Energy with Natural Ventilation

To provide natural ventilation, each room has a low window and an operable high clerestory window on the opposite side of the room. An exhaust fan over the cooking range minimizes heat gain during the summer months. Both summer cooling needs and winter heating requirements benefit from this design, and the designer anticipates that no mechanical cooling will be required in most parts of the Tennessee Valley region.

Energy analyses indicate that this passive solar design will use between 40 percent and 50 percent less energy than a current "super saver" standard of house construction recommended for the TVA region. The optional features of the houses, such as attached sunspaces, may push the energy savings still higher.

Can Solar Make the Difference?

Producing high-quality, energy-efficient housing at affordable cost is the aim of the TVA project. But, sometimes the benefits of passive solar designs may not outweigh the costs. In this case, costs were too high for the builder selected by the TVA. But, the good solar sense of these panel-designed houses could be adapted to a more modest project, perhaps by a custom builder, by a builder/owner or by a production-housing organization. The thermal analyses, the design arrangement and the resulting efficient house form remain a good model for the low-cost solar house.

New Life For A Century-Old Kansas Farmhouse

PROJECT

"The Mossman"
Waugh Residence
Eskridge, Kansas

DESIGNER

Christopher Theis
School of Architecture and Urban Design
University of Kansas
Lawrence, Kansas 66045

BUILDER

Larry Lister
Dover, Kansas 66423

STATUS

Stages One and Two completed, 1982

Estimated Building Energy Performance

BUILDING DATA

Heating degree days	5,182 DD/yr.
Floor area (before addition)	2,070 sq. ft.
Floor area (after addition)	2,930 sq. ft.
Building cost	$75,000
Cost of special energy features	$15,000
Total building UA (before)	1,038 Btu/°F. hr.
(after/day)	961 Btu/°F. hr.
(after/night)	849 Btu/°F. hr.
Heating	
Area of solar heat collector	720 sq. ft.
Solar heat	48 percent
Auxiliary fuel	wood
Auxiliary heat needed:	
(before)	72,750 Btu/sq. ft./yr.
(after)	47,333 Btu/sq. ft./yr.
Cooling	
Cool tubes with low-voltage fans	
Attic fan mounted in cupola	
Auxiliary window air conditioning units	
Cost (before)	17,800 Btu/sq. ft./yr
Cost (after)	no calculation
Hot water	
Need	16.37×10^6 Btu/yr.
Propane fuel cost	$240/yr.
Solar to be installed at final stage	

ANNUAL ENERGY SUMMARY

Nature's contributions	
Heating	66.5×10^6 Btu
Cooling	not calculated
Auxiliary purchased	
Wood	no cost
Electricity	46.4×10^6 Btu
	or 15,836 Btu/sq. ft./yr.

"The Mossman" is a stone farmhouse built in 1884 and located near Eskridge, Kansas, in the tall-grass prairie. The house is occupied by a young couple with three small children—the third generation of occupants. The rural, five-acre site includes a garden, woodlot, apple orchard and tractor barn. It is surrounded by several hundred acres of rolling hills owned by the Waugh family and leased to farmers for pasture and cultivation.

A Wrap-Around Addition

The owner, Curtis Waugh, is a writer and teacher. To design and schedule the remodelling, he hired Christopher Theis, faculty member at the School of Architecture and Urban Design at The University of Kansas. According to Theis, the family wanted, "to retain as much of the exposed stone as possible, which led to the concept of wrapping the addition around the existing structure—exposing much of the stone on the new interior spaces and providing a blanket of insulated space around the mass of the original stone walls."

Specifically, the Mossman renovation consisted of expanding the floor area by roughly 50 percent, adding a greenhouse, living room, study, kitchen, pantry/mudroom, screened porch and new entry porch.

The renovated farmhouse, despite its additions, has not changed into a foreign villa. The additions appear as inevitable developments from the original house, extending and amplifying its native qualities. The resulting form recalls the classic vernacular buildings of Kansas. Stone retaining walls and standing-seam roofs continue the materials and textures of the old house, while insulated frame construction on concrete grade slabs and clapboard siding make it energy efficient.

The concept of the new as a ring around the old is logical aesthetically, thermally and functionally. These new areas act as buffers to winter winds on the north, while on the south they provide welcome sunspaces.

The main entrance to Mossman is off a front porch. The porch is on the west side near the entry drive and large garden, shielding the west stone

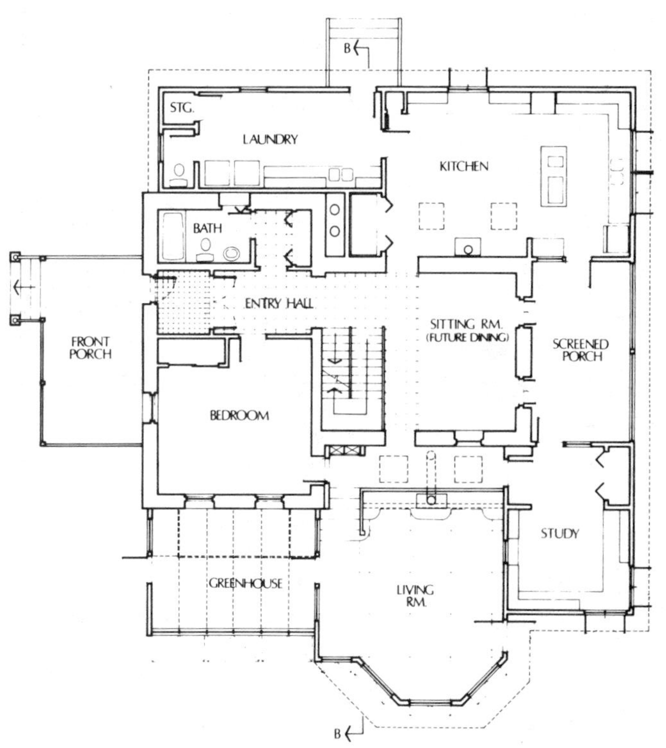

Wider walls indicate original house, with addition wrapped around it.

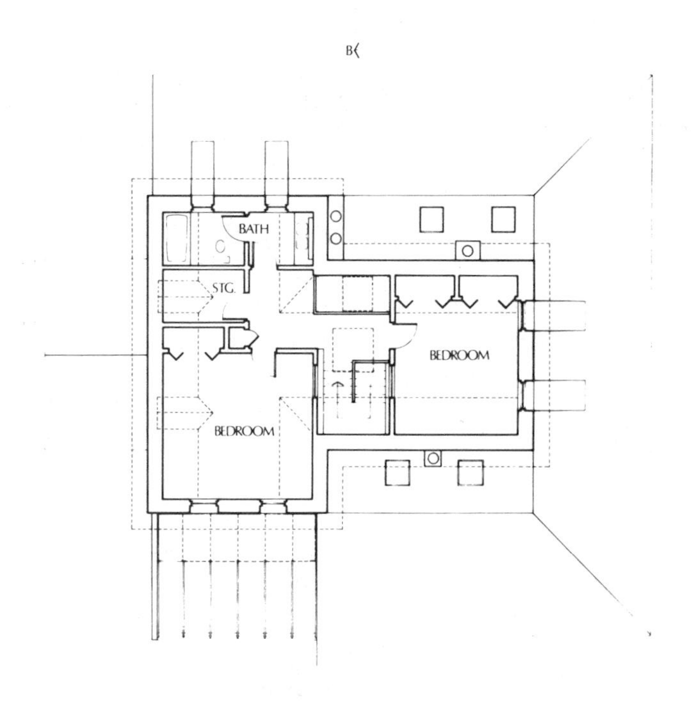

Layout of second story is unchanged from original.

Mossman, before it was rebuilt.

wall from winter winds. The new bath and entry vestibule form an additional barrier.

The daily working entry to the house is from the north into a large pantry that also serves as a laundry and mudroom. With sinks and work counter, the pantry is an ideal summer canning and preserving center; in winter it is a buffer zone where cooler temperatures than those in the living spaces are preferred.

This north side, with its comfortable country kitchen, was the first to be renovated. The large kitchen has 36 feet of work counter, permitting a variety of kitchen tasks. The large space allows informal dining in front of an old-fashioned, wood-burning cookstove that stands against a stone wall. Once the outside of an uninsulated house, the stone now provides thermal mass that reradiates its heat in all directions within the new insulated space. Cooking on a wood stove is part of the Waugh lifestyle, and locating this heat-producing activity on the northeast corner of the house is especially suitable in a passive solar design. Fed for free from the family woodlot, the *Warm Morning* cookstove and the *Defiant* controlled-burn stove in the living room are practically members of the family. The *Monarch* (48,000 Btu/hr.) wood-burning furnace, which is coupled to a forced air gas furnace (140,000 Btu/hr.) for occasional backup heating, is seldom used.

Near the kitchen, and really an informal extension of it, is a sitting room or family room. Since it is in the center of the house, it is always cosy and is also on the path to anywhere else in the house. When the children get older it will become a formal dining room.

Bringing in the Sun for Comfort

The most exciting room, in the way that it ties the old and the new together, is the new living room. Its insulated glass wall with glass roof along the south end of the room is typical of passive solar living spaces. By dropping the floor half a level, the designer was able to enlarge the solar aperture, bringing the sun deeply into the house. This direct solar heat is partially stored in the concrete floor slab and in the exposed stonework. The old stone walls soak up the daytime sun and reradiate their heat back into the adjacent spaces in the evening. The living room is thus something like a solar greenhouse.

The sunspace is ventilated by operable skylights and by the air shaft created by the open stair and cupola that ventilates the entire house. The sunspace glazing is shaded by overhangs in the summer and covered with night insulation thermal shades in winter.

More Passive Collection and Storage

A rock bin below the balcony in the sunspace has two purposes: forced warm-air storage in the winter and cooling and dehumidification of air in the summer. In the past, cooling needs were met completely by window-mounted electric air conditioners. Now, these back up the home's new natural-cooling features.

The adjacent greenhouse uses the mass of stone wall as heat storage and helps to heat the nearby rooms through radiation as well as by direct solar gain. Water-filled drums complete the thermal storage in the greenhouse.

The greenhouse has double-insulated glass and summer shades. It is vented in summer, and the upper panels open out to ventilate the bedrooms on the second floor. *Big Fin* metal shades developed by Steve Baer of Zomeworks provide preheated water to the domestic hot water tank while producing shade in summer. These additions and alterations along the south wall were the second stage in the renovation. They were completed before the final stage—the interior alteration—was begun. Three small bedrooms on the upper floor will be converted to a new bath and two large bedrooms.

Keeping Cool and Conserving Energy

The new and larger stairway opens all three floors to a cupola that contains a large attic fan. The solar-heated air moves up the stairwell in both summer and winter, making it a major source of air movement. For summer, cool ventilation air is induced through new cooling tubes buried beneath the earth. Additional ventilation air comes through low-voltage blowers in the kitchen.

Another place the family keeps cool is the screened-in porch on the east side, a perfect spot for summer meals. It has a view of the orchard and the valley beyond. The screened wall allows natural light to reach the inner windows and helps ventilate the house regardless of the wind direction.

Continuity and Time

In this house, construction materials and methods provide no surprises. Using insulated-frame construction on concrete slabs was the most energy-efficient approach. Stone-faced-retaining walls, clapboard siding, and red composition, shingled roofs continue the materials and textures of the old house, adding both continuity and timelessness.

A Model Remodel

This resourceful passive-solar remodel is a model of the appropriate adaptation of an older house. The logic of every aspect of the redesign and the sensible sequence of the construction work are enviable in their common sense. The extended two-year construction period may seem lengthy, but, when a family intends to live in a house for the indefinite future these months of creating a new house from an old one may not seem insufferable.

Kitchen is roomy and bright.

The project underwent several revisions over the two-year period. The changes, primarily on the south side, came about in response to cost considerations, thermal performance and changes in programmatic needs. The clients wanted a round game table in the new living room and "a place to put the Christmas tree." These needs along with the architect's concern with summer shading and overheating (a major concern in Kansas) resulted in the new shaded bay extension and elimination of roof glazing, which reduced cost and potential problems with shading, insulation and waterproofing. The greenhouse size and configuration changed to enable cost reductions by using standard sections rather than building a custom unit. The west wall of the greenhouse has changed from a glass wall to an insulated frame wall to balance the west elevation, reduce summer overheating and provide better thermal performance in winter.

The effects of this design on energy conservation are impressive. The old house had an annual energy appetite of 72,750 Btu per square foot. The new house will need 47,333 Btu per square foot, of which almost half will be contributed by the sun. That leaves about 25,000 Btu per square foot to be added in fuels, primarily wood from the family woodlot. In sum, the Waughs doubled their space, but their fuel needs dropped by two-thirds.

Sun Up: New Energy For An Old Minnesota Home

RETROFIT RESIDENCE FOR

Jay M. and Kirsten Johnson
Excelsior, Minnesota

DESIGNER

Jay M. Johnson
247 First St.
Excelsior, Minnesota 55331

BUILDER

Peter Boyer
19685 Old Excelsior Blvd.
Excelsior, Minnesota 55331

STATUS

Completed, Winter 1980

Actual Building Energy Performance

BUILDING DATA

Heating degree days	8,382 DD/yr.
Total floor area including basement	2,190 sq. ft.
Building cost	$27,000
Total building UA (day)	419 Btu/ °F. hr.
Total building UA (night)	202 Btu/ °F. hr.
Heating	
Area of solar-heat-collection glazing	236 sq. ft.
Solar heat	17 percent
Primary fuel	natural gas
Heat needed	23,500 Btu/sq. ft./yr.
Cost of heat	13½¢/sq. ft./yr.
Cost heat (previously)	18.8¢/sq. ft./yr.
Cooling	
Not significant	
Hot water	
Not included	

ANNUAL ENERGY SUMMARY

Nature's contributions	
Heating	15.2×10^6 Btu
Auxiliary purchased	
Gas approximately	51.4×10^6 Btu
Total fuel (gas and wood purchased)	30,000 Btu/sq. ft.

SUMMARY

1980–81 (through March)	
Gas	59,100 CF for 6,418 degree days
Wood	1.25 cord oak/birch
1981–82 (through February)	
Gas	62,300 CF for 6,096 degree days
Wood	1 cord oak

Not every older house is worth recycling, but many houses that are less than ideal for retrofitting are not ready to be razed or abandoned. This 70-year old home near Minneapolis, a design exercise in solar rebuilding by architect/owner Jay M. Johnson, illustrates a viable alternative. The project demonstrates how a number of problems common to older, single-family homes in villages or city neighborhoods can be addressed economically, to provide more comfortable and more energy-efficient housing.

Johnson's rebuilding project was an energy-generating addition, a particular response to a growing family's need for more living space, coupled with a continuing commitment to energy-use reduction. Aside from the sentiment of staying in their old house, this solution was an economical alternative to either finding a larger house or adding a conventional addition. Both of these choices would have resulted in larger fuel bills for space heating.

Although the existing house was reasonably energy efficient for its age, it had several drawbacks. The rear kitchen-porch area looked shabby. Roof leaks had caused structural deterioration. And the kitchen was inefficient and less than desirable for family needs. Johnson's rebuilding project not only resolved these problems efficiently, it contributed both thermal energy and quality to the home.

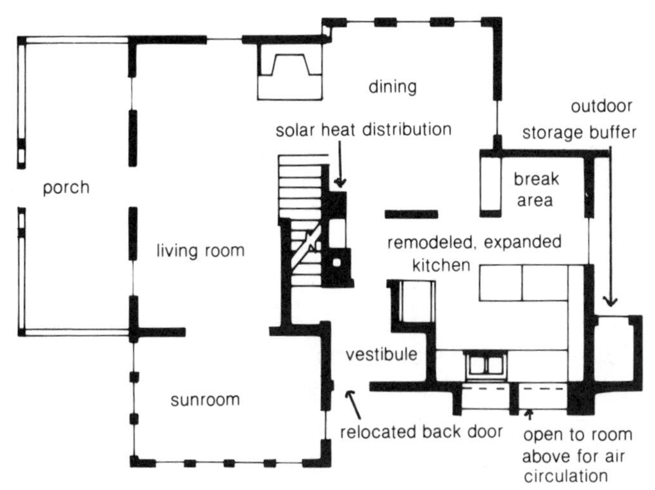

New are entry and expanded kitchen, on first floor.

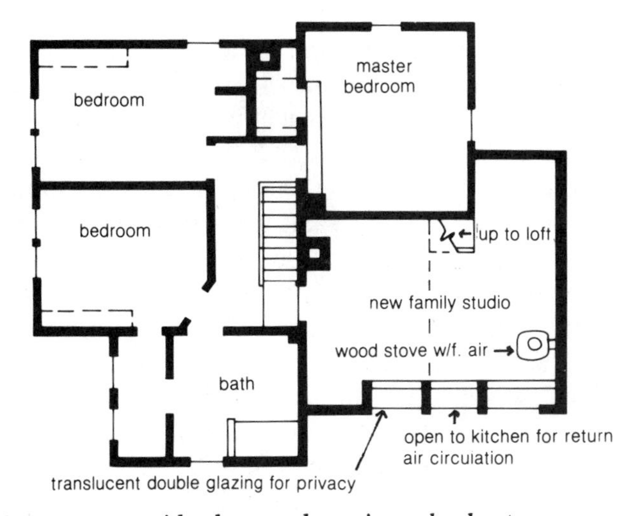

Sunroom provides heat and passive solar heat.

Expanding Upward to the Sun

The Johnson home had several constraints and site problems that presented challenges to an architect planning a passive solar addition. The home was located in a neighborhood of older, closely-spaced houses (six to eight units per acre), situated in a village grid system. In mid-winter, the home was shaded on the lower level; frequently, the "sunroom" was often the "shaderoom." The space above the kitchen, however, received sun much of the day, so this was the logical place for expansion upward to "catch" the sun.

The result was a new second-floor room that serves as a family studio, and on the ground floor, a fully-remodeled kitchen and a new rear entry. The studio has 92 square feet of south-southeast glazing and a south wall of 36 square feet of glass. Together with existing windows, the house now has a total of 180 square feet of south-southeast glazing, 36 square feet of south glazing, and 66 square feet of west-southwest glass—all direct solar gain. The heat is used directly to meet immediate home needs. Due to the modest size of the total solar aperture in relation to the heating requirements, no thermal storage other than an existing brick chimney was needed.

The new kitchen is open to the studio above at the collector wall. A convection loop is set up as air that is heated by the sun rises to the top of the space. It is returned to other areas of the house by the thermostatically controlled, forced-air return of the furnace system.

The upper studio wall is angled to create a nearly south-facing series of clerestory windows (for summer ventilation), which separate a small loft (where the air return is located) and an outside roof deck. The result is a series of open, bright, sunny spaces that substantially reduce dependence on natural gas for heating.

Non-South Building Orientation

As with many towns and cities, the village street grid in Excelsior, Minnesota is not north-south. This house fronts the street approximately 56° west of south, with the side wall 34° east of south. The geometry of the existing house, the close proximity of neighbor's houses, and the minimal expansion possibilities of the small lot essentially eliminated the possibility of a south-oriented addition or wall modification.

Assuming, however, an 85 percent efficient solar gain situation on the 34° side wall and 65 percent on the street side, it remained feasible to utilize the most southerly wall for solar gain, albeit with less benefit. Then, within the confines of a design based on the existing building, some south wall has been generated by developing a roof deck underneath the planes of the roof.

Energy-Efficient Construction Standards for the Infill Addition

In addition to solar gain, energy is captured by wrapping a well-insulated envelope around the existing irregular house volume, reducing the high surface to volume ratio of the old house. Old poorly-insulated walls are thus enveloped by thermally-improved walls of less area, reducing convection losses in two ways. New walls are 2 × 6 studs

This studio has 92 square feet of glazing on the south-southeast side plus a south wall with 36 square feet of glass.

and 6-inch insulation batts with 1-inch styrofoam sheathing and cedar siding; and the roof is constructed of 2 × 12s, with gypsum board over urethane insulation inside and R-30 fiberglass between joists.

The air infiltration of this portion of the house is also reduced due to much tighter construction. The often-used rear entry has been relocated from the northeast to an air-lock southeast entry vestibule.

Convertible Space and Night Insulation

With the resulting higher interior spaces created for both energy and amenity reasons, the convection

This area provides a secluded retreat as well as housing the air return for the heating system.

heat losses are relatively greater than might have been the case with an equally insulated conventional space with minimal windows and 8-foot ceilings. This problem is resolved by making the loft a convertible space for use during sunny winter days and closing it off at night.

Johnson used insulated shades and proposed an intriguing design innovation to reduce convection heat losses, which are substantial in this high-ceiling, multiple-window solar addition. He designed a 4-foot wide hinged panel to close off the solar loft when not in use, thereby restricting upward movement of warm air. He also installed interior 'roman' shades having insulation, fabric, a fabric liner and a vapor barrier. These, he estimates, reduce window heat losses by 15 percent.

Architectural Synthesis

Implicit in any well-conceived addition or retrofit in an older neighborhood is a concern for the architectural character of the area. The Johnsons have done much to ensure that their home does not jar architecturally with the earlier-period homes of Excelsior. Special characteristics of their old home, with its painted cedar shake siding, strong window trim, multi-paned windows, crown-molding trim and gable forms, have been used in new ways in the addition to create a continuity of building materials, color and scale. The impact of the more contemporary, angled form of the south wall/roof deck is somewhat reduced because of the roof shape.

The useable floor area of the house including the 350-square-foot basement was increased from 1,727 to 2,190 square feet so that the family has 27 percent more floor area for living. The new spaces are unique, flexible and sunny in contrast to the more compartmentalized older portions of the house. The exterior has been improved considerably in appearance, and the value of the house has been increased.

Evaluating the Performance of the Renewed House

Johnson estimated that the retrofitted house would use 32 percent less fossil fuel while the floor area would be increased by 27 percent. In fact, in 1981, the Johnson's natural gas budget plan was reduced 23 percent to $27 per month at the same time local newspapers were lamenting the expected 20 percent rise in gas bills. With a combined use of natural gas, wood in an airtight stove and solar, this house has performed at 3.6 to 3.7 Btu/square feet/degree day for two heating seasons. This is a bit higher than the projected 2.8. However, it remains well below the previous load of 5.2 Btu/square feet/degree day. Installing the convertible loft closure and fitting all windows with night insulation should help to bring the Btu figure down near 2.8.

A Passive Retrofit In Minneapolis

Estimated Annual Building Energy Performance

BUILDING DATA

Heating degree days	8,195 DD/yr.
Floor area	2,050 sq. ft.
Cost of special passive features	$18,000
Total building UA (day)	557 Btu/°F. hr.
Total building UA (night)	392 Btu/°F. hr.

Heating

Area of solar-heat-collection glazing	220 sq. ft.
Solar heat	25 percent
Auxiliary heat	gas/hydronic
Cost of auxiliary heat	12¢/sq.ft./yr.

Cooling

Not necessary.

Hot water

Conventional gas-fired domestic water heat. Solar collector planned for future

ANNUAL ENERGY SUMMARY

Nature's contributions

Heating	30.0×10^6 Btu
Cooling	not needed
Hot water	to be installed

Auxiliary purchased

Heating (gas)	92.0×10^6
Cooling	not needed
Hot water (gas)	42×10^6

Total gas purchased

134 MCF @ 1,000 Btu/MCF = 134×10^6 Btu, or 65,350 Btu/sq. ft. for space and domestic-water heating, or 44,900 Btu/sq. ft. for heating only.

SUMMARY OF NATURAL GAS CONSUMPTION

	Million cubic feet of gas to boiler MCF	*Heating Degree Days HDD*	*Cubic feet of gas per degree day CF/DD*	*Btu per degree day per square foot BTU/DD/SF**
1976–77 (*before*)	246	7,626	31	15
1977–78 (*before*)	218	8,516	26	13
1978–79 (*post weatherize*)	121	8,597	14	7
1979–80 (*post retrofit*)	102	7,729	13	6.5
1980–81	97	7,269	13	6.5
1981–82	92	8,400**	11	5.5

*Assumes 1,000 Btu/CF and 2000 square feet heated area.
**7,440 through 3/31/82; assumes 960 through 7/31/82.
Total Gas For 80–81 Heating Season (Boiler off 4/18/82) (Boiler on 10/24/81).

	Boiler Only	*Total Gas Use*
December 31	= 7,764 CCF	2,949 CCF
January 1	= 6,845 CCF	1,824 CCF
Total	919 CCF	1,170 CCF

PROJECT

Pfister House Retrofit
Minneapolis, Minnesota

STATUS

Completed, March 1980

DESIGNER

Peter Pfister, A.I.A.
Architectural Alliance
400 Clifton Avenue South
Minneapolis, Minnesota 55403

BUILDER

Owner built with assistance of
Rocon Construction Co.
1103 Homer St.
Saint Paul, Minnesota 55116

CONSULTANTS

Martin R. Lunde
Structural, Mechanical, Electrical Engineering
Wesley Temple Building
Minneapolis, Minnesota 55403

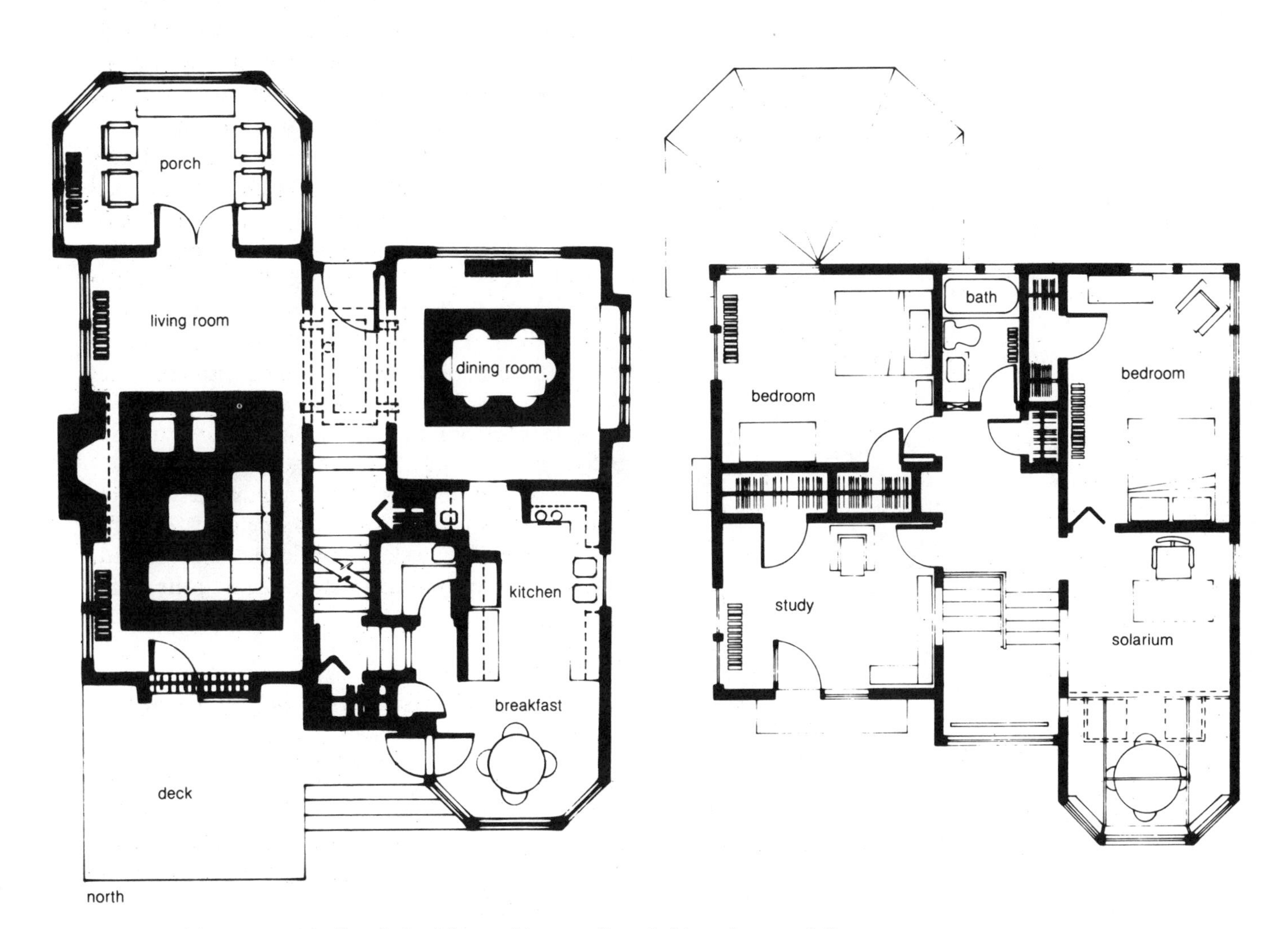

Renovated house now is flooded with sunshine on first (left) and second floors.

Before, south side admitted little daylight.

There are tens of thousands of two-story homes like the 1920 residence purchased by Peter and Darlene Pfister in 1978. Theirs is different, though; they renovated an ordinary, bland, 60-year-old house into a fresh, high performer for a new age. Located in an established neighborhood within five miles of downtown Minneapolis on a 50- × 150-foot lot, the Pfister retrofit was a natural for Peter, who heads the energy division of a large architectural firm.

The existing house was traditional in design, with a central entrance and stairway. Major living spaces were located on the first floor and four bedrooms were on the second floor. The rooms were large and the structure sound and well cared for. More important, the house had a north-south orientation, with room to expand. Major window areas faced north, east and west, so the primary concern was to change the focus of the house from the street (north) side to the garden (south) side, while admitting more daylight and maintaining privacy. The Pfisters' energy goal in the passive retrofit was to achieve a 40 percent annual solar heating contribution that would, with energy-conservation measures, reduce overall space heating needs by about two thirds.

The work done by the Pfisters from 1978 to 1980 did accomplish this goal: The annual natural gas cost for heating the 2,050-square-foot house in this 8,200-degree-day climate is about $250 when achieved. Equally important, however, has been the transformation of a solid, but dark antique into a bright, joyful and still-sturdy home. Especially noticeable is the sunny lightness of the rooms on the south side, particularly the upstairs solarium. Bringing in the sun has both cut heating costs and made the house a more comfortable and appealing place to live.

Remodeling and Insulating First

In May 1979, they moved in and for a year the Pfisters renovated the house—pulling up carpets of varying shades of green, pulling off layers of wallpaper and using gallons of white paint. By relocating walls, they changed the kitchen from a small, square space to a linear breakfast/eating area.

They also insulated the house. Peter devised a method of blowing polystyrene beads into the wall cavities from the interior, using a large cardboard box and a converted Kirby vacuum cleaner. This method was much easier than drilling through the exterior wall stucco from the outside, and it was cheap—about $500 for the wall beads and the fiberglass insulation that they placed into the attic. They also caulked and weatherstripped, added movable insulation to some of the windows and installed water-saving shower heads.

These measures paid off handsomely by reducing the use of natural gas 45 percent from the previous heating season. While he was pleased, Peter could hardly wait to start opening the south wall to the sun.

A DOE Grant and a HUD Prize

In February 1979, with the help of a few friends, the Pfisters began tearing out parts of the existing south wall. In May, they began the passive retrofit with the help of a grant from the U.S. Department of Energy (DOE).

The DOE grant required the couple to instrument and monitor the house and its use of energy during the 1979–1980 heating season. The instruments included: two photovoltaic pyranometers to record solar radiation; wind-speed and wind-direction indicators; a gas-flow meter to record gas consumption of the boiler; a heat-flux transducer to measure heat flow through various construction elements; and status indicators to monitor the position of the automatic window insulation and the operation of the air-circulation fans.

In addition, the Pfisters installed about 40 thermocouples to measure temperatures in various locations. All the information was fed into a computer located in an open closet in the kitchen. As the house progressed, measurements were recorded. But, they could not say exactly how well it worked because the 1979–1980 heating season came and went without all the renovations completed.

Besides the DOE grant, Peter's preliminary plans for the house received a U.S. Department of Housing and Urban Development (HUD) award in the Passive Solar Design Competition. Thus encouraged, the Pfisters plunged ahead on the passive solar south wall.

A (Mostly) Passive Heating System

The existing heating system was a gas-fired boiler with a gravity hydronic radiator distribution system. This system did not cause air to be recirculated, so distributing the passive solar heat was a primary concern.

The retrofit consisted of adding two primary solar-gain areas and two secondary solar-gain areas, increasing the south-facing glass from 50 to 220 square feet. A 9- × 12-foot, two-story solarium was added to the kitchen/breakfast area at the southeast corner of the building. This sunspace allows direct gain into the building. Quarry tiles in the kitchen floor absorb and store the sun's heat and radiate it into the room when the temperature drops. Warm air rises through the metal grate flooring to the top of the two-story solarium where it circulates to the north side and interior of the house via the "cutouts" in adjacent walls. The metal grate floor above allows low-angle winter sunlight to filter deep into the kitchen space. And the openings in the walls between the solarium, stairway and the northeast bedroom and the removal of the top of the wall between the solarium and the master bedroom give a feeling of openness and a sharing of light rare in houses of this period. With the help of a few open windows on each side of the house, these openings keep the house cool and breezy during the summer.

Now, rooms on south side are sunny and bright.

The second major passive-gain area is a full-height window bay in the stairway landing leading to the second floor. The daytime sun is augmented by external reflectors. The area behind the glass contains 25 phase-change, thermal storage "energy rods," each 3½ inches in diameter and 6 feet long, that absorb radiant energy during the day to control overheating and release it at night. A small (¼ horsepower) fan draws warm stratified air from the ceiling of the central stairway and returns it to the north side of the living room and dining room. To complete the loop, another small fan draws air off the floor of the living room and returns it to the stairway bay thermal-storage area.

The two secondary passive-gain areas are incorporated into the south wall of the living room and the bedroom above.

Solarizing A Ranch-Style Home

RESIDENCE REMODELING OF

Elliot/Mulready House
Belmont, California

DESIGNER

Roger East
745 Emerson Street
Palo Alto, California 94301

SOLAR CONSULTANT

Michael Riordon
514 Bryant Street
Palo Alto, California 94301

STATUS

Unbuilt

Estimated Building Energy Performance

Suburban California Retrofit

BUILDING DATA

Heating degree days	2,800 DD/yr.
Cooling degree days	108 CDD/yr.
Net floor area	2,383 sq. ft.
Building cost	$100,000
Cost of special energy features	$7,000
Total building UA	1,037 Btu/°F. hr.
Heating	
Area of solar heat collection	263 sq. ft.
Solar heat	72 percent
Auxiliary heat	natural gas
Auxiliary heat needed	52,469 Btu/sq. ft./yr.
Cost of auxiliary heat	15¢/sq. ft./yr.
Cooling	
Natural convection via solar chimney	
Auxiliary fans for rockbed	
Cost of auxiliary cooling	$400
Hot water	
Need	14,000 gallons/yr. @ 120°F. (incoming 60°F.)
Auxiliary natural gas cost	$18/yr.

ANNUAL ENERGY SUMMARY

Nature's contributions	
Heating	114.0×10^6 Btu
Cooling	not calculated
Hot water	25.2×10^6 Btu
Auxiliary purchased	
Gas	28.6×10^6 Btu or 12,000 Btu/sq. ft.

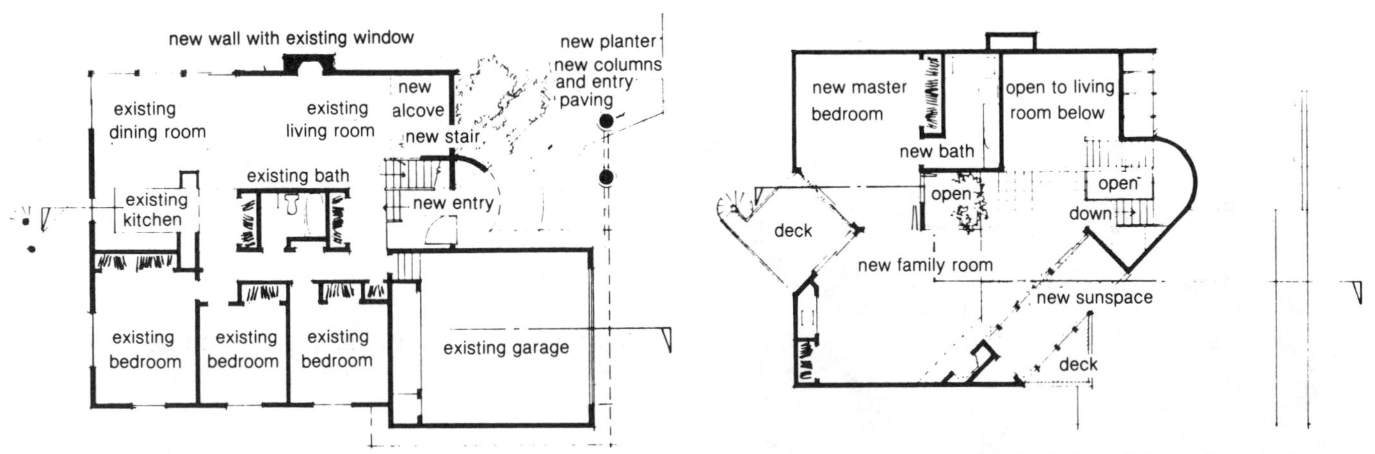

Design creates new alcove and entry on first floor (left). Floor space is nearly doubled with addition of second floor (right).

The Elliot/Mulready house is a 15-year-old, wood-frame, post-and-beam house in a pleasant, settled neighborhood in Belmont, California. Its stained shingles and exposed-beam ceiling are clearly part of the California ranch style that has been so widely copied elsewhere. The family would like to add a new, large entertainment/family room to the one-story dwelling but without an "add-on look." They also want a means of passive solar heating and an active solar domestic hot water system. Architect Roger East's design solution will nearly double the amount of floor space and dramatically change the exterior and interior feel of the house.

There are some obstacles to this metamorphosis. The house is sited at 45° off true south—not an auspicious beginning for a solar retrofit. And, because the house is one story, additional space can only be created by adding a second story—but the existing flat ceiling roof is too light structurally to carry the load.

To solve the latter problem, a new floor will have to be built over the existing roof. And for the former, the new sunspace on the second floor will have to be placed on the diagonal to be oriented to true south. This dramatic new angle and the rest of the new second story will transform the low California look into a more three-dimensional villa. The solar chimney and extended flue pipes will give the house a new skyline. Not only will the family gain ideal solar access, they will also enjoy spectacular views to the north from the upstairs windows and deck.

Moving Warm Air and Saving Energy

Inside, portions of the existing roof/ceiling will be opened up to give a two-story volume over part of the living room. Yet the existing ceiling beams of the first floor will be kept intact. A vertical opening around the stair and the new entry, and another vertical opening in the center of the house will tie together the two floors. These openings will provide routes for air to move freely, convecting interior warmth or coolness.

Another aspect of the remodeling, though less obvious, will be an energy-conservation program. To help reduce the heating load, about 40 square feet of single glazing will be removed from the existing windows and replaced with double-glazed window units. All new windows will be double glazed; and all existing doors will be recaulked and weatherstripped. Fortunately, the original walls were insulated with R-11 fiberglass batts in the initial construction. These will not be changed. However, the existing roof will be entirely covered with the new addition, which will have R-19 walls and R-30 roof insulation.

The existing house, a conventional ranch style.

A Changeable (and not so Balmy) Climate

Why all this insulation for balmy California? In fact, not all of California has a balmy climate; in Belmont, the heating load is 3000 degree days. Also, the daily weather is erratic. Fog and unseasonal rains can boost the heating requirement at odd times of the year. For such a changeable climate, energy conservation, passive thermal storage and various control devices are desirable in any solar application.

About the Solar Calculations

Because of his familiarity with the quirks of the Bay Area climate and with the realities of domestic solar applications, solar consultant Michael Riordon was asked to calculate the performance of the Elliot/Mulready retrofit. Riordon predicted that the passive system devised for the house would provide 72 percent of the annual heating requirement. His predictions and his short method of arriving at performance figures for the house are given in Appendix C.

How the Passive Solar System Works

Because the new space in the Elliot/Mulready house is to be on the second story, architect East and consultant Riordon rejected Trombe walls, water columns, and roof ponds as unfeasible due to their weight. Instead, since the house had no basement, they designed a 2-foot-deep thermal storage rock bed in the crawl space under the first floor to store the heat from the sunspace-heated air. Conduction and convection will distribute the heat into the living space when needed.

Warmed air from the sunspace will be ducted to the basement using an in-line, 1,500-cfm fan. There, the warm air will vertically charge the 1,000 cubic feet of 2-inch river rock from the top down. Warm air will be discharged from the bottom of the rock bed through a 4-inch perforated drain pipe, collected by a larger manifold and returned to the sunspace through another duct. Backdraft dampers will prevent reverse thermosyphoning during the cooler evening hours. The rock bed was estimated to contribute 38 percent of the solar heat.

The warm air will also be conducted through the insulated wood floor to heat the interior spaces, and floor registers can be opened to admit more heat when needed. Two-story spaces in the living

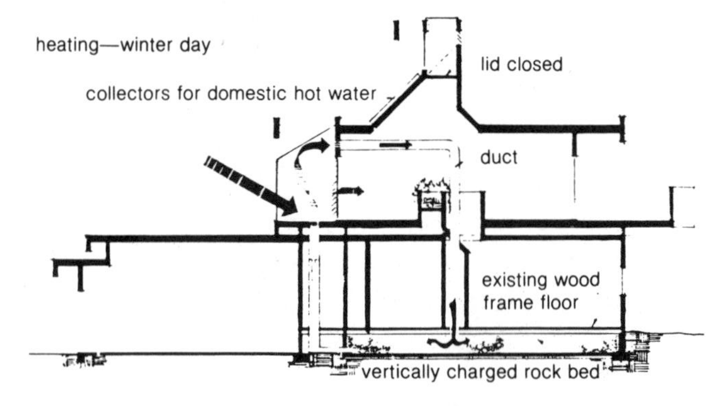

Heat from sunspace is spread throughout house.

room will carry the warm air to the second story, and Casablanca-style ceiling fans will prevent temperature stratification.

Additional direct solar heat gain will come from the southeast and southwest windows (19 percent) and from the window wall between the sunspace and the living space (15 percent). Dark-colored venetian blinds on the sunspace side of the window wall will control the amount of sun admitted.

Cooling with a Solar Chimney

The primary cooling system will be the solar chimney that exhausts warm air from the entire house. During the summer months, the family can open the insulated door in the chimney, allowing hot air to rise. A mat of expanded metal mesh, painted black, will be located in the chimney behind glass to absorb heat; it will add heat to the rising air and accelerate the natural air flow. The family can adjust the amount of warm air that is exhausted by using operable vents.

At night, the supply fan will draw cool air into the rock bed and exhaust it through the sunspace vent. During the day, air will be circulated through the cool rocks to keep house temperatures down.

Shading from the summer sun will be provided by overhangs and porch roofs, and, in the sunspace, by roll-out shades.

A Good System, But Not Yet Built

House costs in California are high, and construction is expensive. The estimated cost for the Elliot/Mulready renovation in 1980 was $100,000, of which only $7,000 could be allocated to energy-conservation measures, even including the passive solar system. As a result, construction of the Elliot/Mulready retrofit has been postponed to await more favorable times and lower interest rates.

Nonetheless, the potential metamorphosis of this residence from a woody California speculative builder house to a custom-designed villa is in itself an interesting case study. Aside from the potential 72 percent solar savings is the excitement of the house itself. By virtually doubling the size in adding a second story, the house has developed an entirely new series of perceptual dimensions, both internal and external. The new use of the site, especially by opening up distant views, is dramatic. The transformation to a more international and cosmopolitan architectural expression is equally impressive. This promising design demonstrates how even in the potentially restraining context of remodeling, solar design need not dominate other important and interesting architectural directions.

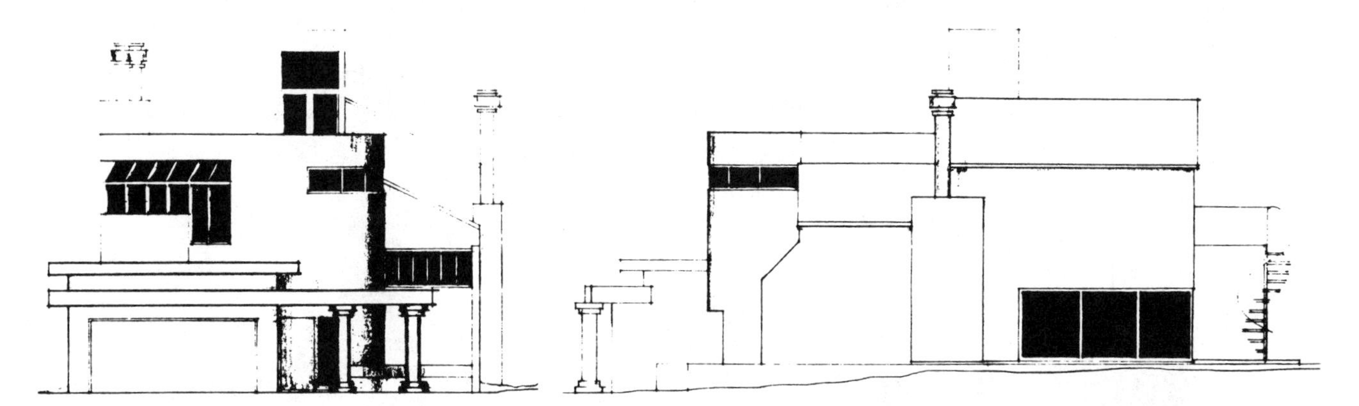

Southeast (left) and northeast views. Note in drawing at left how sunspace is placed on the diagonal so it is oriented to the true south.

A Lively Solar Duplex In Berkeley

PROJECT

Solar Duplex
3035 Colby Street
Berkeley, California 94705

DESIGNER

David Baker
1950 Bonita Street
Berkeley, California 94704

BUILDER

David Baker

STATUS

Occupied, November 1978

Estimated Annual Building Energy Performance

BUILDING DATA

Heating degree days	2,909 DD/yr.
Floor area	1,200 sq. ft.
Property cost	$28,000
Renovation cost	$48,000
Cost of special energy features	$9,000
Total building UA	686 Btu/ °F. hr.
Heating	
Area of solar-heat collection glazing	286 sq. ft.
Solar heat	90 percent
Auxiliary, apartment 1	⅛ cord of wood
Auxiliary, apartment 2	750-watt electric baseboard
Cost of electric auxiliary heat	4¢/sq. ft./yr.
Cooling	
Stack effect natural ventilation	
No auxiliary	
Hot water	
Need	21,900 gallons/yr. at 125 °F. (incoming 55 °F.)
Solar-collector area	
Solar fraction	70 percent
Auxiliary electricity cost	$47/yr.

ANNUAL ENERGY SUMMARY

Nature's contributions	
Heating	47.9×10^6 Btu
Cooling	0.4×10^6 Btu
Hot water	8.3×10^6 Btu
Auxiliary purchased	
Heating	8.2×10^6 Btu
Cooling	none
Hot water	3.5×10^6 Btu
Total electricity (thermal only)	11.7×10^6 Btu or 4,035 Btu/sq. ft.

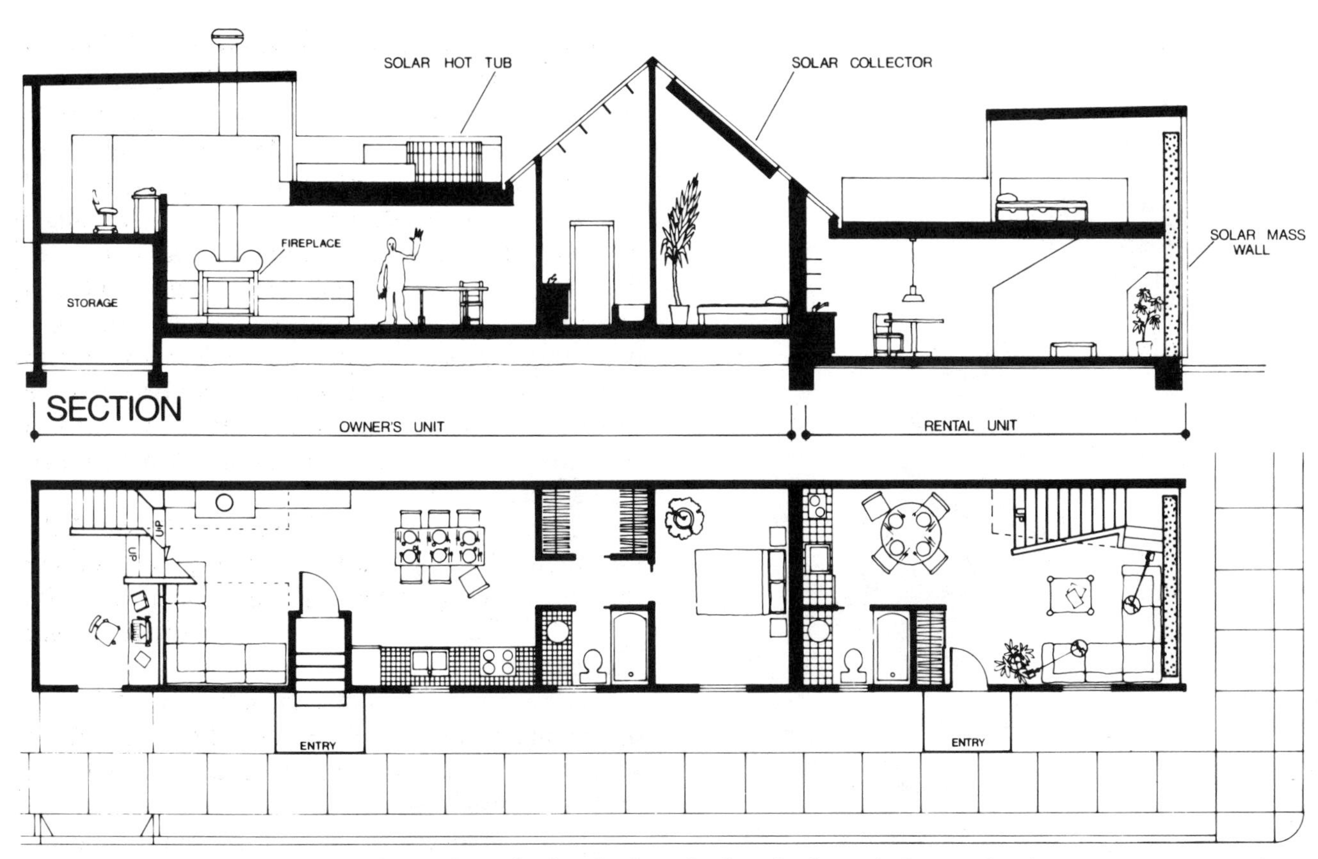

Sectional drawing (above) and floor plan of solar duplex. Both units have decks—and privacy.

Originally built in 1911 as a corner grocery, and later used as a Chinese laundry, this urban solar duplex had become a rundown set of decrepit apartments by the time it was purchased in 1977. David Baker, an architecture graduate from the University of California at Berkeley, redesigned and rebuilt the structure for his own home, which he now calls "Palazzo Wyatt Earp." And he refers to the architectural style as "Post Old West." His new design demonstrates how to make a solar purse out of a sow's ear.

The energy wisdom of this urban renovation goes well beyond its skylights and Trombe walls. Renovation in itself is a highly resource-conscious ethic. And in a fully built city, upgrading existing structures is a very good investment, particularly when you get to live in the transformed space. There are also professional advantages to a solar architect's living in a solar house of his own design.

Baker's solar duplex has received wide recognition. Perhaps its stylishly scruffy appearance, complete with a California hot tub, means it may never be duplicated. But, its modest size and charming details make it both photogenic and nonthreatening. It looks small, friendly and simple. Its design message is that retrofits can have personality, even eccentricity. How lively their eccentricity is depends on the creativity of the designer and not the restraints of solar access or zoning regulations.

A Lot and a Building with Problems

The corner lot in Berkeley measures 78 feet long on east and west and 14 feet wide on north and south—hardly favorable dimensions for solar orientation. In fact, the complete 1,092-square-foot property is covered by a building that is 1 inch wider than the lot!

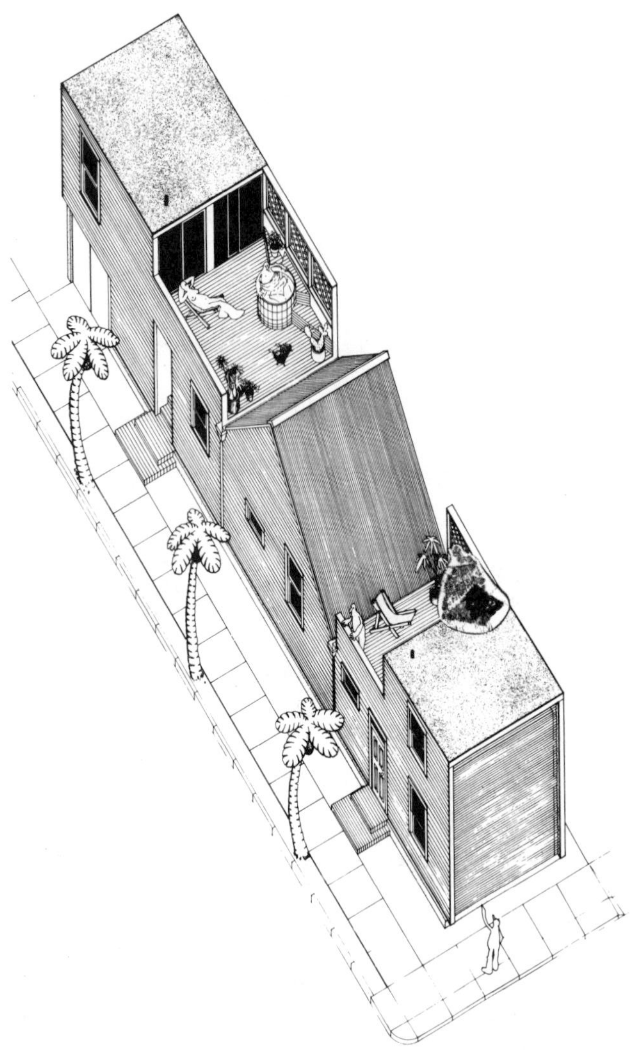

Bird's-eye view shows how roof effectively separates two deck areas.

The existing patchwork building had serious structural defects and building-code violations, yet because of zoning restrictions, any new construction had to fit inside its dimensions. Fortunately, a 3-foot-high attic and tall parapets allowed a new second story to be tucked into the building without increasing the apparent size from the outside. This avoided the complex process required by the city to gain approval for a new building.

The circulation pattern of the original plan was inefficient and inconvenient. The narrow lot forced a building that consisted of a single row of rooms, lined up so that one had to pass through the rooms in sequence to get anywhere. Windows opening onto the sidewalk created a conflict between the need for privacy and for light and air.

The final design responded to several problems and opportunities in addition to passive and active solar-energy concerns. Frequently, single elements of the design solved more than one problem. For example, the roof area between the two apartments was raised into a gable to provide a mounting surface for the solar collectors; it also serves as a visual and acoustic barrier between the private roof decks. And these decks, one with a solar hot tub, solve the problem of no outdoor private space.

Strips of skylights on the gable roof bring light into the entire length of the building, overcoming the impossibility of having windows on opposite sides of the rooms and permitting additional solar heat gain.

Double-height rooms, including the 8- × 8-foot living room, visually expand the limited floor area and allow hot air to be exhausted through roof vents in the summer.

Other changes were made for livability reasons alone. The outside entrances were relocated to the middle, minimizing traffic going through one room to another. The new plan also used space that would normally be vacant, such as the area between the linear kitchen and eating area.

The Trombe Wall Passive System

A radiant Trombe wall provides over 90 percent of the space heating for the rental apartment. Located at the extreme southern end of the building, it takes up most of the surface area of the wall.

The thermal storage is a steel-reinforced, 8-inch-thick concrete block wall, grouted solid and finished with red-clay, unglazed quarry tile on the interior surface. It is 16 feet high and 12 feet wide and has a thermal storage capacity of 2,800 Btu per degree Fahrenheit change in temperature.

The Trombe wall receives full sunlight for at least six hours every day throughout the winter, making it an ideal solar collection surface. The solar collection element consists of 220 square feet of vertical glazing—in this case corrugated, fiberglass-reinforced, plastic greenhouse glazing with an additional layer of Tedlar film laminated to the weather side. It receives the most solar radiation when the sun is low in the sky to the south, which is also the period of greatest heating demand. In summer, the sun's rays strike the wall for only a portion of the day, and then from an oblique angle, automatically eliminating overheating.

The Trombe wall performs well on a typical sunny day in winter. The interior temperature averages about 20 degrees warmer than the ambient exterior dry-bulb temperature, particularly in the evening when thermal lag causes the concrete to be warmest while outside temperatures are dropping rapidly.

Direct-Gain Solar Heat

The gable roof is oriented due south at a slope of 40 degrees from horizontal and contains active solar collectors for hot water as well as 50 square feet of direct-gain skylights. They extend completely across the roof from east to west to form three continuous light slots. Double-glazed with the same corrugated, fiberglass-reinforced plastic used for the Trombe wall, they bring direct solar gain into the owner's bedroom and into the kitchen and bathroom of the rental unit.

In addition, two aluminum sliding glass doors totaling 75 square feet of double glass are oriented to the south in a clerestory over the living room of the owner's unit. These doors face out onto the deck and provide most of the heat required for the unit. A ceiling fan immediately behind these doors in the top of the double-height living room destratifies the hot air that rises to this point.

A New Slab Floor and Water Tubes

The existing wood floor in the rental unit was replaced with an insulated concrete slab on grade covered with the same red quarry tiles that cover the back of the Trombe wall. Two inches of polystyrene insulation (R-7) was placed on top of the drainage gravel underneath the slab. This new floor stores direct gain from the strip windows framing the Trombe wall and from the skylight over the kitchen/bathroom area.

Because the owner's unit has no structurally integrated thermal mass, two water tubes 10 feet high and 18 inches in diameter were installed in the bedroom. Each holds 120 gallons of water, treated to prevent the growth of algae. Their thermal storage capacity is about the same as the Trombe wall —2,600 Btu per degree Fahrenheit of temperature change. Considering the risk of earthquakes, the builder/designer braced the tubes with steel brackets to restrict lateral movement. More than tripling the thermal mass of the room, the tubes are very effective at moderating temperature swings in the bedroom and keeping it comfortable.

Solar Hot Water System

An active solar, domestic hot water system serves both living units and provides some heat for the rooftop hot tub. Three 4- × 10-foot collectors, with all-copper absorber plates, are integrated with the south-facing gable roof, using the same glazing type as the Trombe wall. This glazing also covers the direct-gain skylights with a single monolithic waterproof surface.

To prevent damage if the collector piping leaks, Baker installed asphalt roll roofing under the absorber panels. Special high-temperature foamboard insulation protects the waterproof membrane from the heat within the collector enclosure.

Hot water from the collectors is stored in two tanks with a total capacity of 200 gallons; a variable-flow differential thermostat controls a small circulating pump that prevents the system from freezing by circulating water for a short while when temperatures fall below 35 ° F.

The active solar system also provides backup heat for the 5-foot-diameter hot tub through a copper coil heat exchanger under the hot tub bench. The remaining tub heat needed is produced by an efficient gas thermosyphon heater.

Brookside: A Two-Family House

PROJECT FOR

Duplex Investors
Yorktown Heights, New York 10598

DESIGNER

William T. Meyer, A.I.A.
Peter Wormser
The Ehrenkrantz Group
19 West 44th Street
New York, New York 10036

STATUS

Unbuilt

Estimated Building Energy Performance

BUILDING DATA

Heating degree days	5,131 DD/yr.
Cooling degree days (70 °F. base)	536 CDD/yr.
Total duplex floor area	3,966 sq. ft.
Building cost	$123,000
Cost of special energy features	$13,000
Total building UA (day)	375 Btu/ °F. hr.
Total building UA (night)	318 Btu/ °F. hr.
Heating	
Area of solar-heat collection glazings	410 sq. ft.
Solar heat	70 percent
Auxiliary heat	electric water-to-air-heat pump
Auxiliary heat needed	6,214 Btu/sq. ft./yr.
Cost of auxiliary heat	7¢/sq. ft./yr.
Cooling	
Natural cross ventilation plus stack ventilation through roof monitors	
Auxiliary type water-to-air heat pump	COP 2.5
Cost of auxiliary cooling	5¢/sq. ft./yr.
Hot water	
Need	58,400 gallons @ 115 °F. (incoming 51 °F.)
Oil cost	$38/yr.

ANNUAL ENERGY SUMMARY

Nature's contributions	
Heating	35×10^6 Btu
Cooling	9×10^6 Btu
Hot water	none
Auxiliary purchased	
Electricity and oil	56.6×10^6 Btu or 14,280Btu/sq. ft.

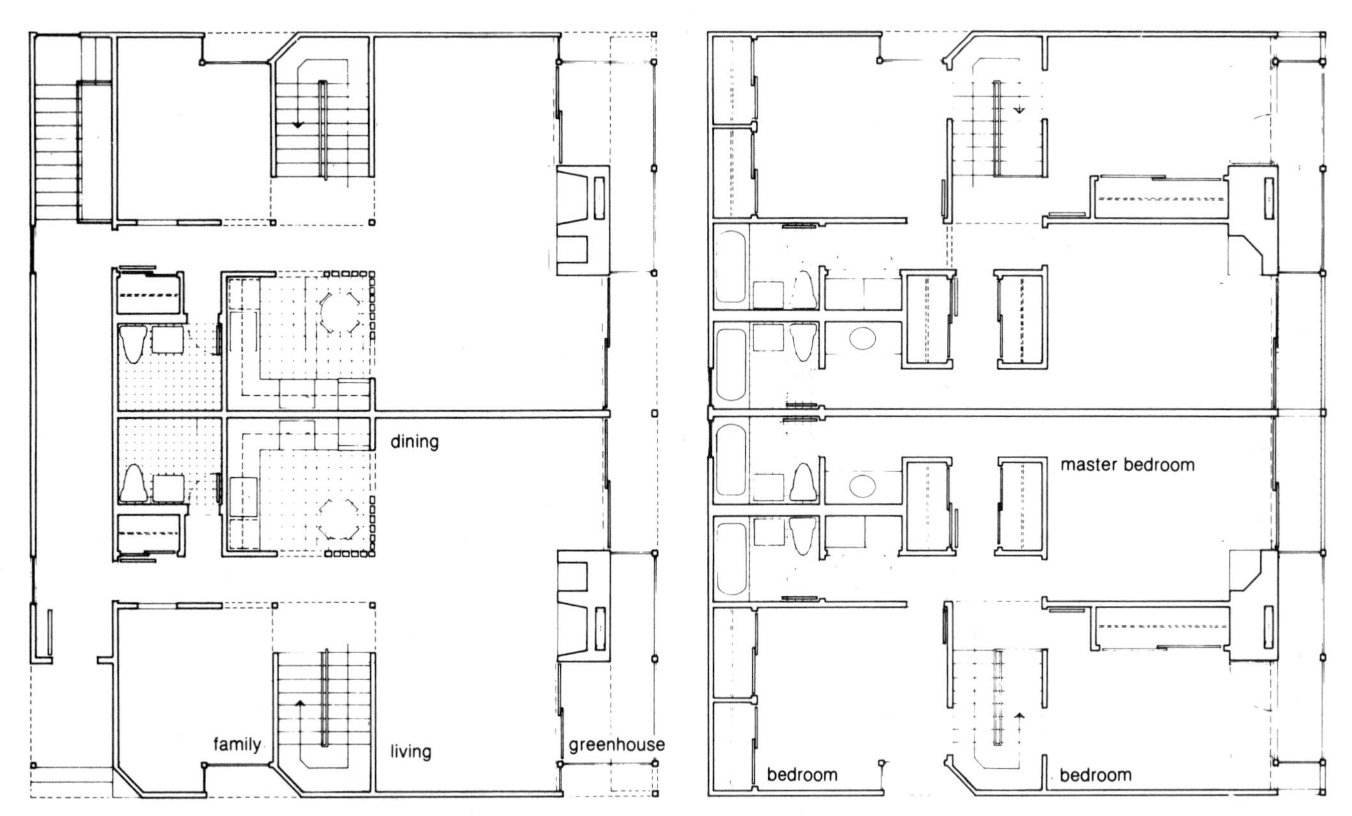

First (left) and second floor plans of duplex. Added greenhouses and windows contribute 70 percent of annual heating requirements.

Brookside is a two-family duplex, a traditional American house type that combines most of the advantages of the single-family dwelling with economy of construction. An efficient building, the duplex can be made even more efficient when it incorporates passive design features. The use of passive solar applications, of course, has become more popular as fuel prices have risen and the cost of land near cities has kept pace. The first priority for the Brookside house, therefore, was a moderate price suitable for a group of small investors.

The two designers of the Brookside house, William T. Meyer and Peter Wormser, are young architects in The Ehrenkrantz Group, a large design office employing 120 people on the top floors of a Manhattan skyscraper. This firm has an international reputation as specialists in high-technology applications and, while they typically work on large buildings, they also have focused on small-project prototypes such as the Brookside project.

A Narrow Lot

The site is a narrow lot on the east side of Saw Mill River Road in Yorktown Heights, Westchester County, New York. The west end of the house faces a busy road, while the south opens on the entry drive to a side yard. The sloped site allows for parking in the basement at the rear.

The two-story house is essentially an elongated cube that faces south. Its architectural interest comes from recessed windows on east and west together with porches, greenhouses and roof monitors facing south. The clapboard exterior and the roof monitors reflect the traditional clapboard buildings common to Westchester towns, with their charming trellised porches and wooden cupolas.

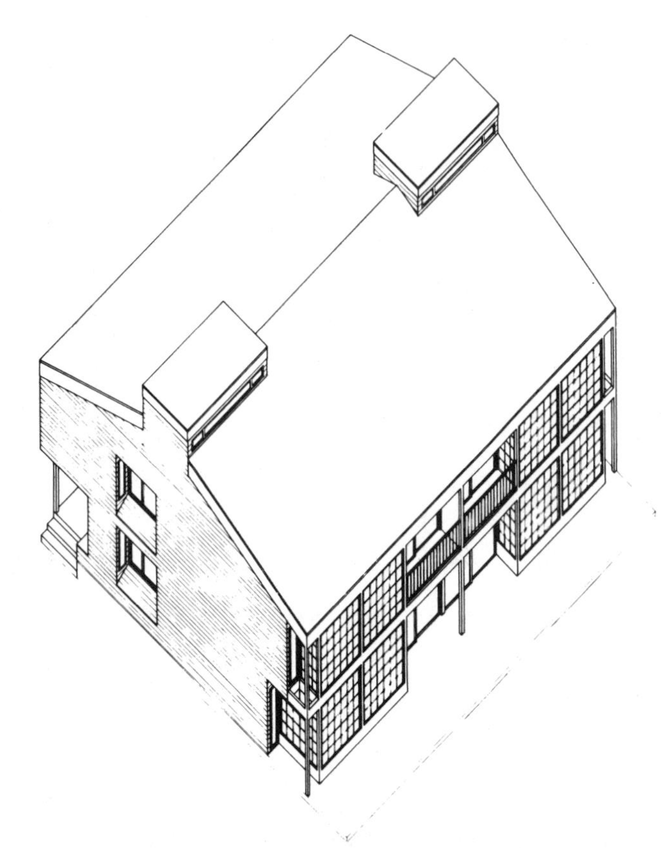

Mass of glass features south side, at right.

Energy-Conserving Construction

The framing for the Brookside house is conventional 2 × 4 studs. Insulation is 3⅝-inch fiberglass batts supplemented with an extra layer of 1-inch, rigid-foam insulation board instead of plywood sheathing. The roof (R-44) is framed with 2 × 12 rafters and insulated between the rafters with fiberglass batts, allowing interior sloped ceilings. Because the ceilings slope down to walls less than the standard 8 feet, the volume of the house is reduced. The foundation is a concrete slab on grade, insulated with 2-inch rigid foam board at the perimeter. Principal openings are on the south side, with smaller and fewer openings on the north, east and west. All openings are double glazed, and a double-glazed, air-lock entry further minimizes any heat losses through infiltration. This energy-efficient construction is now current practice throughout the United States and is wise whether passive solar techniques are used or not.

Mirror Images

The 3,200-square-foot building consists of two living units, each with three bedrooms and two-and-one-half baths. The approach and entrances are from the corners of the north side. These two apartments mirror each other in their interior layout. They are oriented along a north-south axis with principal living spaces on the south to enjoy the greatest benefit from direct winter sunlight. The bedrooms are clustered on the north to provide an insulating buffer against winter winds.

Greenhouses and Glass Doors

The primary passive-solar features are the greenhouses and the glass doors and windows on the south facade of the house. These glass areas contribute approximately 70 percent of the annual heating requirements. Direct solar gain from the low winter sun raises interior temperatures and is absorbed by the concrete floor slab, which radiates the heat back into the living spaces later in the day. The daytime heat gains are retained and nighttime heat losses prevented by insulating curtains and blinds. Indirect solar gain through the greenhouses supplies heated air to each living unit. Warmed air rising from the greenhouses, from the concrete slab and from the fireplace mass, is collected in the roof monitors and returned with fans to the living spaces to prevent extreme temperature stratification.

Cooling With Passive Solar

In the warmer months, passive cooling is achieved by window shading and natural ventilation. In addition to designed cross-ventilation, the roof monitors above the stairwells allow rising hot air to es-

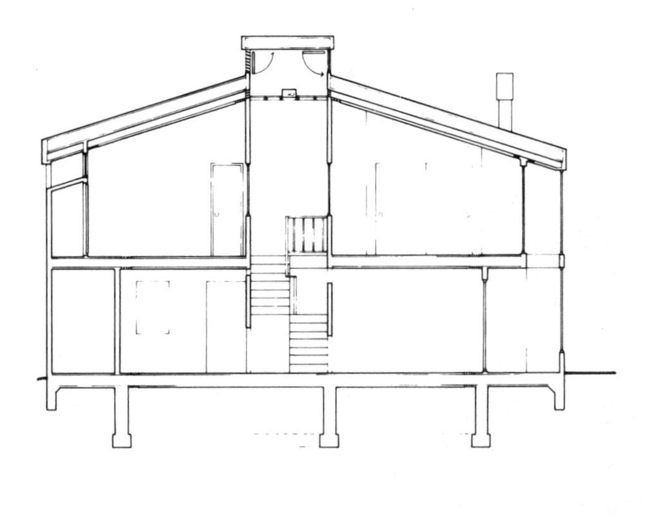

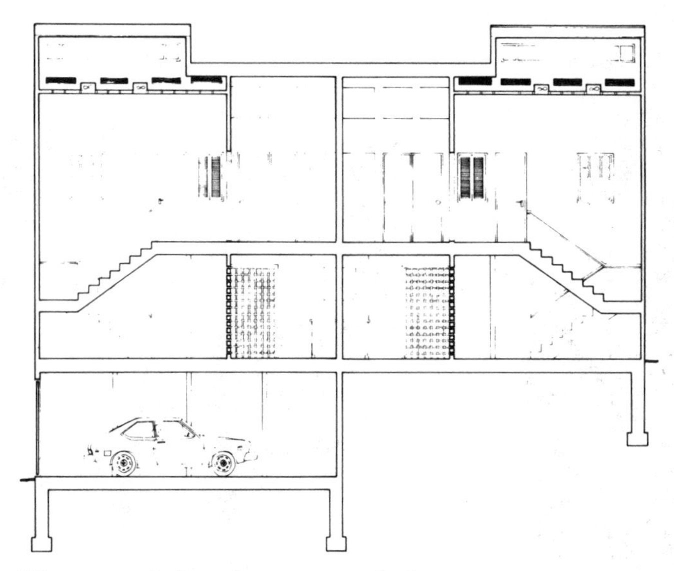

Sectional drawings show roof monitors above stairwells. These permit hot air to escape during summer.

cape, creating a natural stack effect that draws in cooler air from the first-floor openings. Solar penetration on the south side is minimized by calculated overhangs and internal shading devices, such as blinds.

Water-to-Air Heat Pump

During the heating season, the water-to-air heat pump provides auxiliary heating when not enough solar heat is available from the greenhouses. Using cool ground water as its source, the heat pump also helps to cool the house centrally when natural ventilation is insufficient. In addition, a central oil-fired furnace is available to serve as a backup heat source. Costs for auxiliary heating are estimated at $224 per year per unit.

Prototype For Investors

This interesting compact design provides an impressive prototype for housing investors. Its simplicity adds to the economy generally inherent in duplex plans. Nevertheless, the interior is generous without being extravagent and the exterior has character wtihout being outlandish. In the clarity and logic evident throughout, this design provides a prototype that could be adapted to many other sites and a variety of climates.

A Townhouse Design For A Cold Climate

PROJECT FOR

Canadian Townhouse
Kitchener, Ontario

DESIGNER

James Fryett, Project Designer
Joseph C. Somfray, Architect
122a St. Andrew Street West
Fergus, Ontario
Canada N1M IN5

STATUS

Unbuilt

Estimated Building Energy Performance

BUILDING DATA

Heating degree days	7,491 DD/yr.
Unit foundation area	672 sq. ft.
Net unit floor area	1,548 sq. ft.
Total building UA (day)	490 Btu/°F. hr.
Total building UA (night)	286 Btu/°F. hr.
Heating	
Area of solar-heat collection glazings	349 sq. ft. plus 111 sq. ft. sloped at 23.5°
Passive solar heat	75 percent
Night insulation savings	28,200 Btu/sq. ft./yr.
Auxiliary heat	active solar air plus in-duct electric resistance heater

Cooling

Passive overheating controls
Night flushing plus thermal mass

Hot water

Solar collectors inside greenhouse
Sizing not determined.

ANNUAL ENERGY SUMMARY

Nature's contributions	
Heating	32×10^6 Btu
Cooling	not calculated
Hot water	not calculated
Auxiliary purchased	
Electricity and wood	11.6×10^6 Btu or 7,500 Btu/sq. ft.

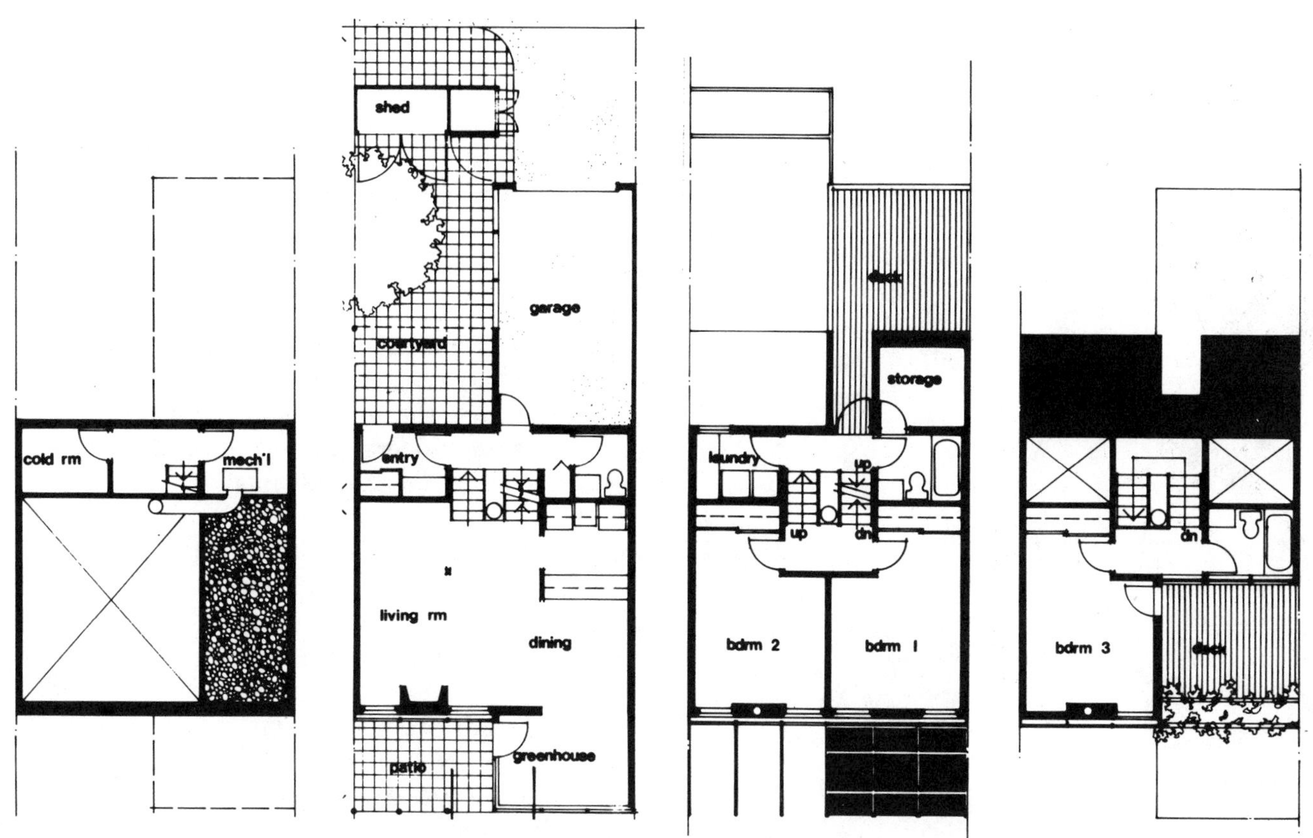

Four of the six levels of this house include, from left, cold room area in basement, first floor, second floor, and third floor.

Typical rowhouses or townhouses are energy conservative, land efficient and resource sensitive, and solar townhouses can be even more so. James Fryett, the designer of this Canadian solar townhouse, has used active solar collection as an auxiliary to what is basically a passive solar building. His intent was to design a comprehensive system that would integrate the many strategies necessary to respond to a cold climate. Fryett describes in his own words his design concepts and their synthesis:

"This townhouse design is an attempt to achieve a high level of passive heating for the cold climate of southern Ontario. The design for a steep, south-facing, sloped site in Kitchener, Ontario is intended as a generic solution. The intent is to realize the densities permitted by local zoning criteria while maintaining appropriate levels of public and individual identity. Identification and separation of various zones, such as private and public outdoor areas, vehicular and pedestrian routes, play and service areas is deemed important to the scheme as a living environment. Site planning considerations include winter winds, summer shading and solar rights."

Complexity and Directness in a Severe Climate

The more severe the winter climate, the more complex will be the design of components to control natural energy flows. Here, many different strategies are integrated within a generic design that can be repeated. Because the problem is not unique, the solution can be adapted or modified to fit many similar situations.

Successful solar designs for severe climates are always interesting. Here, for instance, every south-facing surface is a solar heat or light collector of

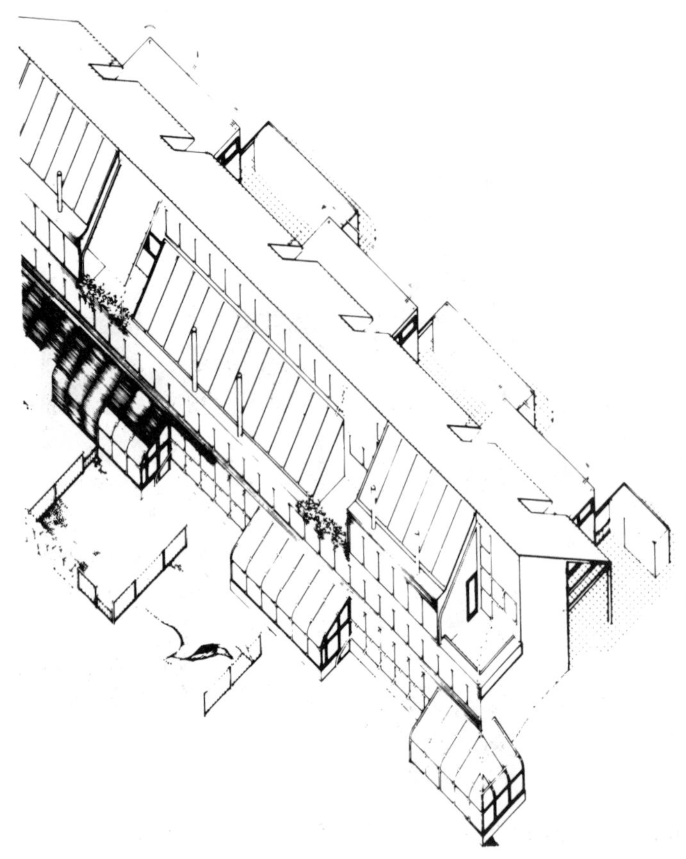

An array of glass brightens south side of these townhouses.

one type or another. A three-dimensional house that has six different living levels is one way of reaching for the sun. And at a latitude of 43.5° N., the sun at winter solstice is low enough in the sky to encourage such a vertical solar architectural design.

A vertical arrangement is always economical in land use, and the quality of living space in a vertical townhouse is always exciting. Each room and space has some special relationship to the outdoors and to the other parts of the house. There are no dark corridors and no wasted spaces. Plumbing runs and heating ducts are short and direct. While the design appears complex, each component has been fitted into the whole with admirable clarity and directness. Although this particular design has not been built, it has strongly influenced several similar projects that have.

Energy Conservation in the Form and Fabric of the Building

The basic design strategy is to reduce heating requirements by conserving energy in the form and fabric of the building. These conservation techniques and passive devices are synthesized in an architectural form that is visually strong and unifying. The random operation of control devices does not detract from its aesthetic quality.

The entire building envelope, including floor slabs and foundation walls, is insulated. All glazing is sealed double glass. Infiltration and conductive losses through the building skin are limited on the noncollecting surfaces; a continuous layer of 1½-inch foam insulation with epoxy-stucco finish over all 2 × 6 studs prevents thermal bridging. All sill plates and corner posts are caulked, and a continuous 6-mil. vapor barrier surrounds the heated spaces. The building volume has been reduced by limiting floor-to-floor heights to 8 feet. This is achieved by using open fir joists and 1½-inch tongue-and-groove flooring. Night insulation for all windows and Trombe walls is manually controlled by the occupants.

In addition, a substantial conservation of energy is achieved through zoning various spaces and activities. The air-lock entry, unheated storage area and garage on the north side of each unit protect heated spaces from cold winds, as do the courtyard and north roof. Further protection will come from coniferous planting in the courtyard areas. The stairway/circulation space, laundry, and washrooms are all internal secondary buffers.

Solar Heat and Building Construction

Passive techniques for heat collection include both direct and indirect solar gain. On the lower level, the south-facing greenhouse admits solar gain on winter days, with thermal storage provided by the floor slab and back wall. The water-filled Trombe wall on the next level provides both thermal storage and a natural thermocirculation loop, which carries heated air up the stairwell. On the top level, the high sunspace also admits direct gain. Heat is stored in the floor slab on the upper level. As temperatures become excessive at the top of the sunspace, hot air is ducted by using a fan to the rock bed at the lowest level, where it warms the floor slab on the lowest living level. Heat from the rock bed is also cycled through the air-distribution system.

Passive solar gains cannot, however, provide all the annual heating requirements for the units. Thus an active solar air-distribution system has been provided for as backup, with an electric heater built into the duct. For domestic hot water heating, collector fins can be mounted inside the greenhouse, as can extra water tanks or eutectic salt tubes for additional heat storage. As a heating backup in each housing unit, there is a patented wood-burning fireplace stove that uses outside combustion air.

To save building costs, the design calls for conventional building materials and techniques. The basic building module is a 4-foot square grid. All water-loaded Trombe wall components can be premanufactured off site. Building framing is wood stud and standard insulating materials are used. Other cost- and energy-saving features are common walls and shared service stacks.

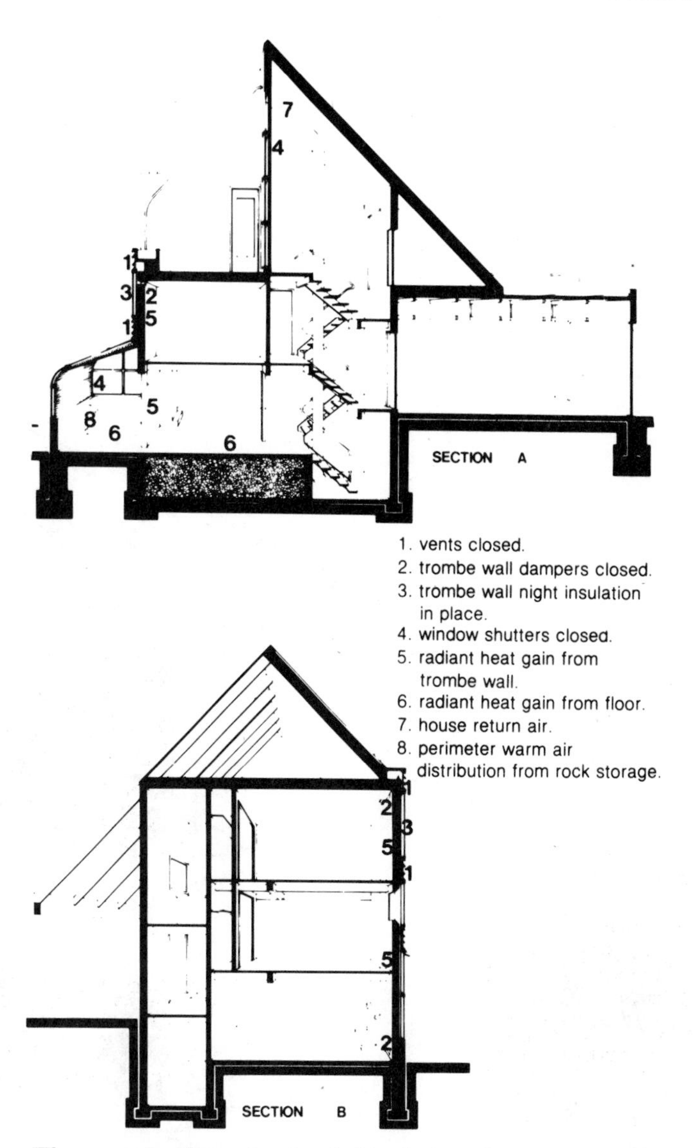

The complex floor levels of this design. In upper drawing, the left side faces south, in lower drawing, the left side faces north.

Moving Air for Winter Heating and Summer Cooling

In general, heat flows vertically upward at the building perimeter, during heating periods; and air returns downward through the stairwell. During the day, this loop operates naturally until high temperatures are reached in the sunspace at the top. At that point, a differential thermostat control directs the heated air to the rock storage bed. The active air collectors mounted on the roof are also coupled with the air-recovery duct and its controller. With the air-recovery duct, the stairwell prevents warm air stratification and stimulates air movement in an otherwise draft-free space.

While high outdoor summertime temperatures are rare, building overheating can be a problem. Here, comfort is maintained by manipulating night insulation on the Trombe wall and shading devices on the greenhouse and sunspace. Careful planting of deciduous vegetation on the south side will also help. Air flow through the building is enhanced by venting the Trombe walls and opening north vents. The thermal mass of the building also helps moderate temperature swings.

Condominiums On A Tight California Site

PROJECT

Melrose Condominiums
Glendale, California

DESIGNER

William R. Watkins, Jr.
while associated with
Robert Zachry, Zachry Associates
2300 Westwood Boulevard
Suite 10
Los Angeles, California 90064

STRUCTURAL ENGINEER

E. Brad Graves
1700 Westwood Boulevard
Los Angeles, California 90024

STATUS

Unbuilt

Estimated Building Energy Performance

BUILDING DATA

Heating degree days	2,015 DD/yr.
Floor area	1,487 sq. ft./unit
Building cost	$725,000 for all seven units
Cost of special energy features	$1,450/unit
Total building UA (day)	351 Btu/°F. hr./unit
Total building UA (night)	325 Btu/°F. hr./unit
Heating	
Area of solar-heat-collection glazing	112 sq. ft./unit
Solar heat	99 percent
Auxiliary heat	resistance electric
Auxiliary heat needed	4,624 Btu/sq. ft./yr.
Cost of auxiliary heat	2¢/sq. ft./yr.
Cooling	
Natural convection	
Auxiliary cooling	electric fans
Auxiliary cooling needed	344.8 Btu/sq. ft./yr.
Cost of auxiliary cooling	15/100¢/sq. ft./yr.

ANNUAL ENERGY SUMMARY

Nature's contributions	
Heating	68.1×10^6 Btu
Cooling	not calculated
Solar hot water	to be fitted after sale
Auxiliary purchased	
Electricity	5.25×10^6 Btu or 35,305 Btu/sq. ft.

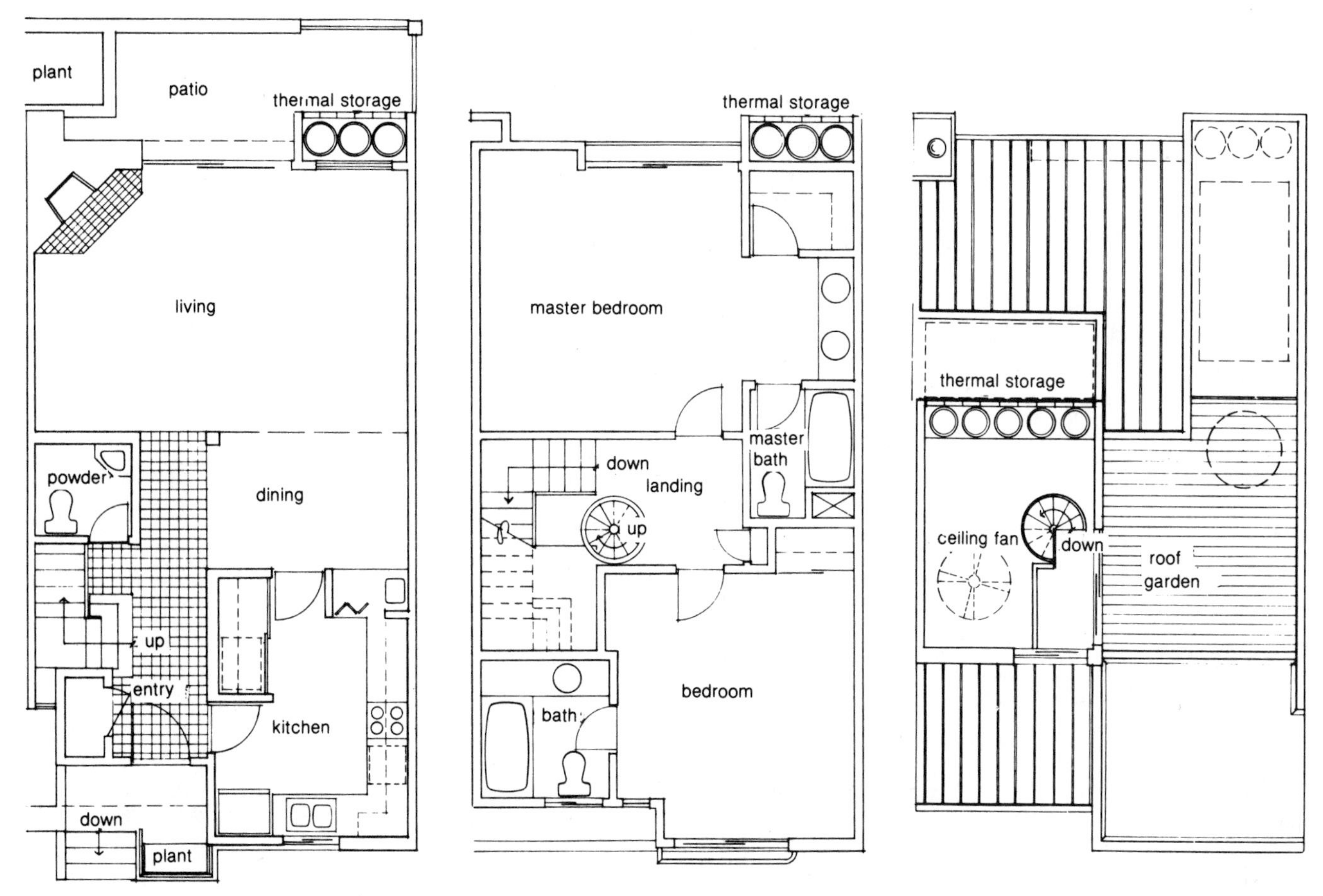

From left are the first, second, and roof garden levels of this condominium.

This sensible seven-unit condominium design belongs well within the genre of apartment buildings in the Los Angeles area. Its insulated wood-frame structure and stucco-finish surfaces are appropriate to the southern California climate. Built-in parking is necessary both to get planning approval and to accommodate residents. The exceptional aspects of the project are a synthesis of its good planning, its lively appearance and its potential thermal performance. It is not a design that is likely to stop traffic. But, as a competent model of maximizing land use and optimizing passive principles, this condominium is a fine example of a sound design.

The design was prepared for a local developer/contractor who specializes in small, high-density condominium projects. Market analysis of the area indicated a need for moderate-sized, energy-efficient, two-bedroom units. And, to fulfill this need, two-story townhouses with common walls appeared to be the obvious choice. Unit land cost made it imperative to "build out" the site, or use it most efficiently. Local zoning and condominium ordinances indicated a maximum of seven units would be allowed if the designer placed the townhouses over a subterranean garage.

Moving from Private to Public to Private Spaces

These constraints formed the parameters of a preliminary design. The approach to the seven housing units is a common concourse—a raised walk along the north edge of the property. To enhance the

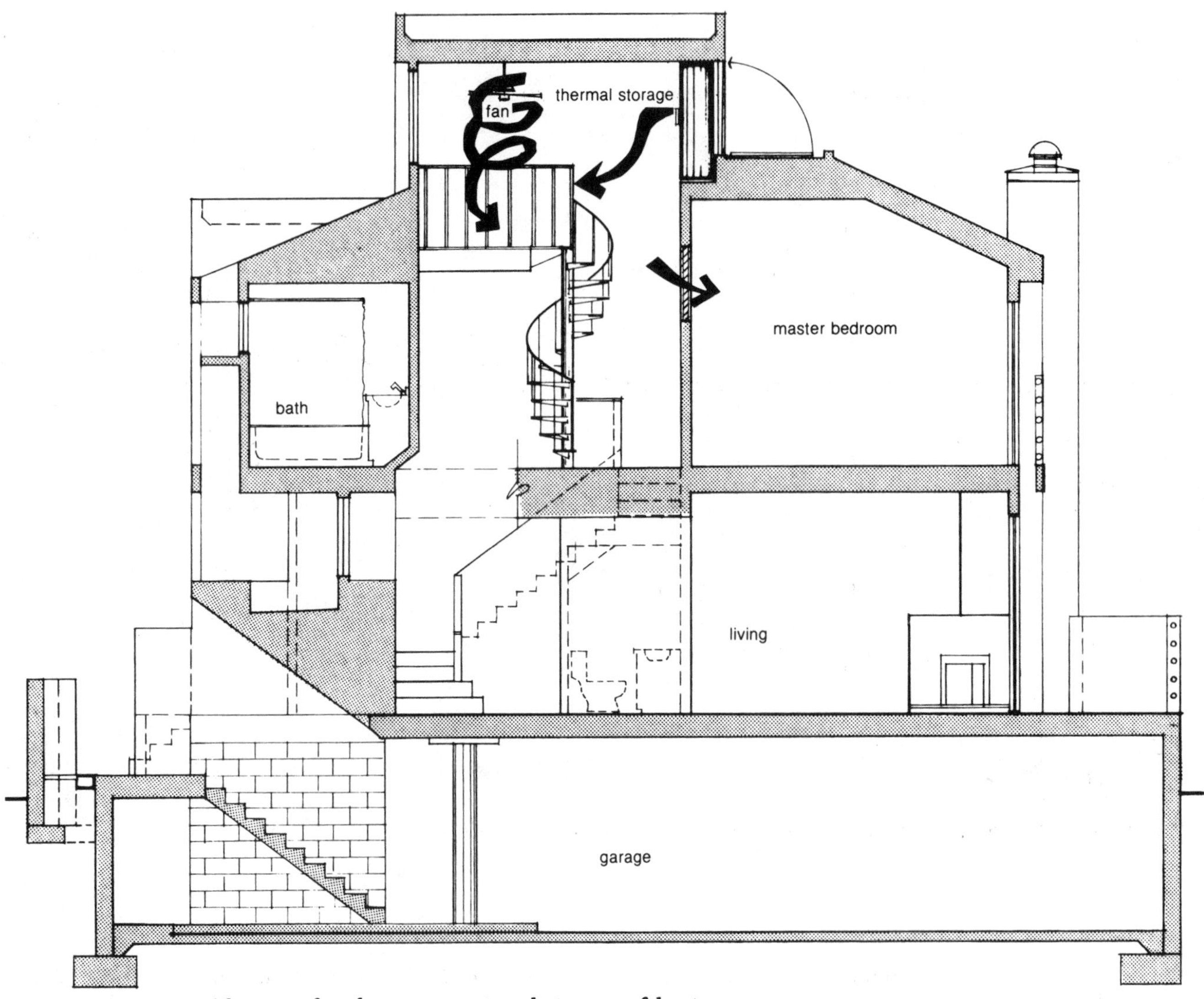

Circular core provides area for the movement and storage of heat.

community aspects of the project, the design was planned around a sequence of private to public to private spaces. The resulting architectural form reflects this sequence.

Thus stairs were provided for each two sets of parking spaces in the garage level, one stair serving two units. These stairs lead to the more public concourse, which in turn leads to a short flight of stairs at the entry patio of each unit and finally to the entrance. Planters at the entrance of each unit enhance its individuality within the project.

In contrast, the south side provides a series of private patios in addition to the solar heat collection areas and ventilation openings.

Organizing Around a Circulation Core

The interior of each housing unit is organized around a circulation core, with emphasis on light and texture to define space. Tiled flooring at each level articulates the circulation core. The three-story space leading to the roof garden is designed as a vertical transition that uses light and form to link the three levels. It brings in the sun and ties together the building elements vertically, both functionally and environmentally.

Passive Applications in a Moderate Climate

Southern California does not have a severe climate by comparison to other areas of the country. The average number of heating degree days is about 2,015 per year. But the normal passive applications in this moderate climate must satisfy cooling as well as heating needs.

The primary purpose of a passive heating system in this area is to provide enough stored energy to sustain comfort with temperatures of 30° to 50° F. during the night, an occurrence not only in the winter but also in the summer. This storage capacity must be joined with a system for ventilation and exhausting of hot air during the summer days. Flexibility, therefore, becomes a major requirement.

Designing the Heat Collection Systems

Because of the density requirement and the configuration of the property, there was a limited area available in which to place the solar heat collection surfaces. This limitation was made worse by the need for some cross-ventilation and a concern for privacy from the property owners next door. The final configuration was a combination glazing for direct gain (window) and indirect gain.

The indirect-gain portion is a two-story, stacked "water wall" closet, separated from the living space for control. Heat is absorbed in fiberglass cylinders filled with water. Warm air is drawn off the top of the solar closet and distributed to the living space with low-velocity (1,500 cfm) duct fans. Replacement air is drawn in through a louvered duct in the living room wall, thus completing a convection loop. The bedrooms are also fitted with louvered panels leading into the stairwell for circulation of air even when the bedroom door is closed.

The direct-gain system consists of south windows in the living room and master bedroom as well as a water wall in the upper portion of the stairwell tower. This portion is fitted with an insulating, manually operated exterior shutter to be used in times of cool weather to prevent heat loss. Air is distributed from the warmed water mass by a reversible-pitch, 3,500-cfm ceiling fan located at the top of the stairway tower.

Cooling with a Heat Sink

Hot-weather cooling is by direct heat transfer to a heat sink—a 13-inch concrete slab floor above the garage—and by cross-ventilation and the stack (venturi) effect. The tremendous volume of the stairwell tower, with 53 square feet of operable windows at the top, gives a good air-circulation exhaust path. Other energy-conserving features such as double-insulated glass also help to keep the house comfortable in both heating and cooling seasons.

The Value of Unbuilt Projects

Every architect has had the experience of commissioned and speculative designs which are not constructed. The reasons are infinite and often have nothing to do with the quality of the design. Such is the case in this instance, where the developer chose not to execute the design due to a declining market for condominiums resulting from high interest rates. But many projects which do not continue to fruition serve to attract interested new clients and become a useful tool for securing new commissions. Most important, each project, built or unbuilt, contributes toward its pool of resources and continuum of expertise.

Windcreek Condominiums

PROJECT FOR

Coldiron and Cobb
Sacramento, California

DESIGNER

Mogavero + Unruh
Architecture and Development
811 J. Street
Sacramento, California 95814

COMPUTER MODELING

Davis Alternate Technology Associates
Davis, California 95616

STATUS

Unbuilt

Estimated Building Energy Performance

BUILDING DATA

Heating degree days	3,143 DD/yr.
Cooling degree days	1,228 CDD/yr.
Floor area (per unit)	1,140 sq. ft.
Building cost	$37,000/unit
Project cost	$1,100,000
Cost of special energy features	$1,400 net/unit
Total building UA	421.8 Btu/°F. hr.
Heating	
Area of solar-heat collection	139 sq. ft.
Solar heat	64 percent/yr.
Auxiliary heat	two wall-type gas furnaces
Auxiliary heat needed	7,456 Btu/sq. ft./yr.
Cost of auxiliary heat	5¢/sq. ft./yr.
Hot water	
Cost	$108 electricity/yr.

Units will be plumbed for solar use
Larger (120-gallon) tank included in construction

ANNUAL ENERGY SUMMARY

Nature's contributions	
Heating	15.1×10^6 Btu
Cooling	16.3×10^6 Btu
Hot water	(to be installed later)
Auxiliary purchased	
Electricity and gas	23.3×10^6 Btu or 20,438 Btu/sq. ft.

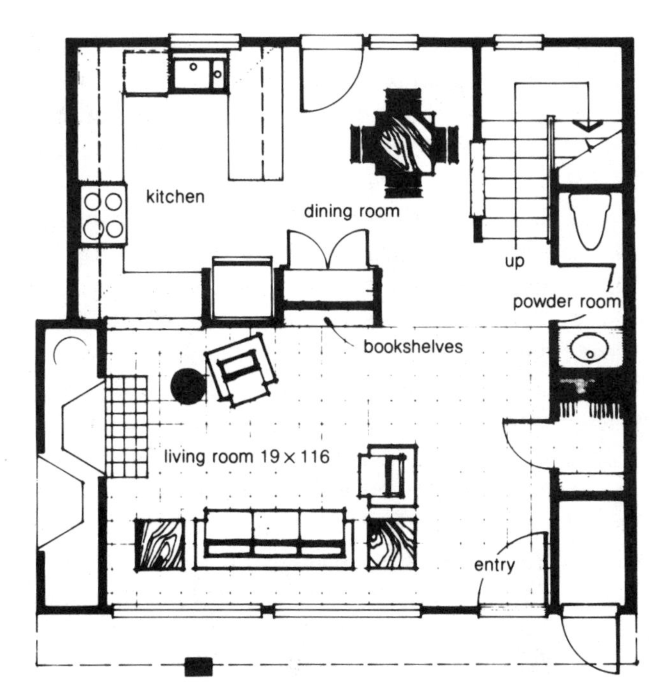

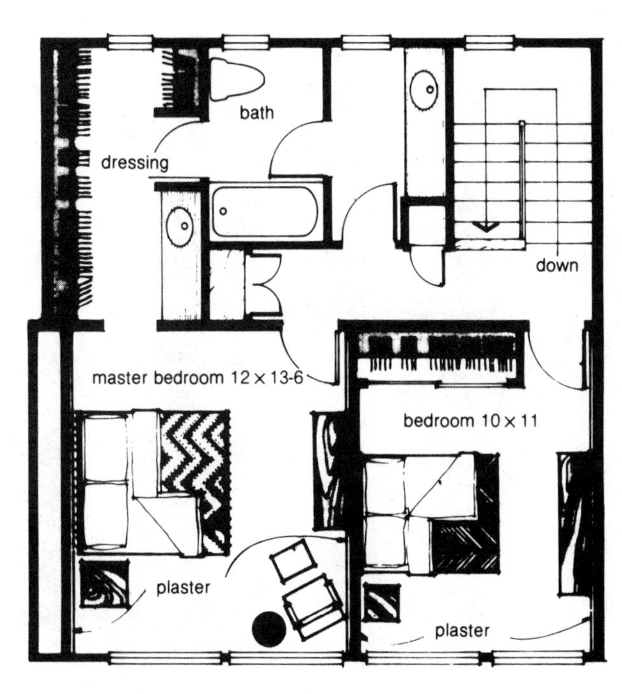

First (left) and second floor plans of Windcreek Condominiums.

Designing Windcreek Condominiums presented several special challenges to the architects, David Mogavero and J. Ronald Unruh of Sacramento, California. First, they had to meet the developer's requirement for easily affordable units. This placed a premium on economy in size and layout. Also, the design had to incorporate passive solar techniques for Sacramento, a place with hot summers and rainy, overcast winters. Careful site planning and landscaping were required to maximize both the amenities and the yield of the land. Although Windcreek has not yet been built, the resulting design has become a prototype for speculative, economical, multi-family developments. Many of the concepts described here have been adapted in projects built by Mogavero and Unruh.

The Windcreek Site Plan

The site design calls for eight one-bedroom flats and 32 townhouse units with suitable paved parking and access areas. Five different unit types are clustered together as neighborhoods, with front doors oriented to pedestrian "streets" in clearly defined groups. The sense of groups is reinforced by lattice archways at points of entry and by paving and landscaping changes. Lattice work also screens grade walls and shades the south elevations.

Landscaping plans at Windcreek include large-specimen shade trees around parking areas, along the west boundary and in open spaces for summertime cooling. Paved areas are kept to a minimum and are well shaded. The longest walk to a front door from the parking area is 40 to 45 steps—about 100 feet. Every townhouse unit has a minimum of a 12- × 24-foot backyard, and some yards (for the three-bedroom units) are as large as 28 × 24, almost the size of backyards in single-family tract houses. For maximum privacy and openness, most yards run along the property line.

Attention has been paid in the site planning to such miscellaneous needs as laundry rooms and garbage enclosures. Provision has been made for such amenities as a pool, a spa, a play area and play equipment for children.

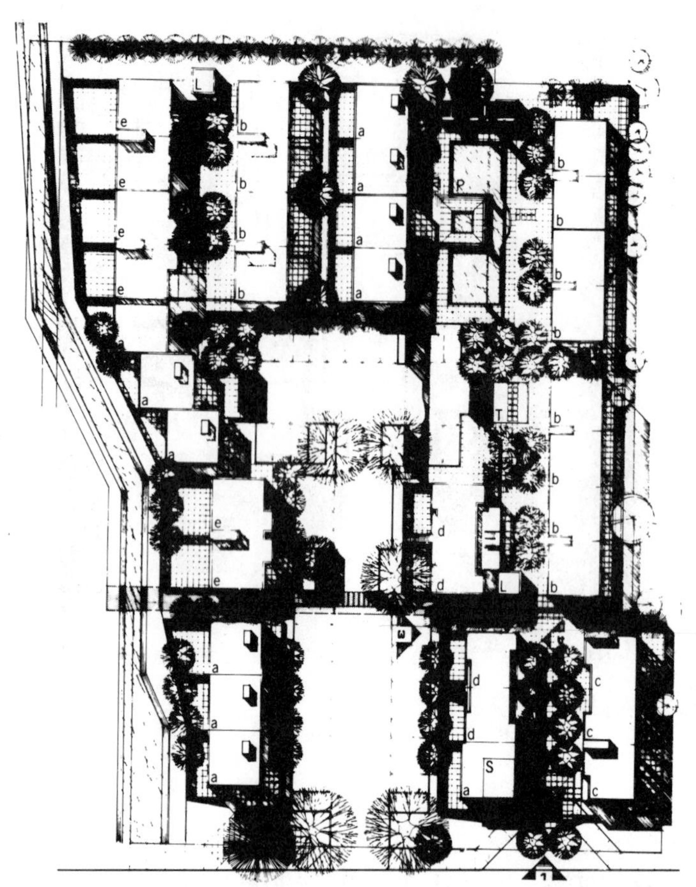

Site design provides open spaces for summertime cooling.

Unit Layouts and Plans

The architects compared the economics of wide units versus narrow units and found that the common wisdom of the economy of narrow units with a long depth was not necessarily true. An almost square unit plan allows a much broader solar access. More important in an earthquake zone, the lateral shear bracing necessary could be handled by solids of the south window wall as well as the parallel interior walls.

The townhouses share common east-west walls, with 24 feet from unit to unit. The almost-square configuration allows for five different apartment sizes and layouts, with the two-story units being the most efficient. The sizes are modest, ranging from one-bedroom units (602 square feet) to two-bedroom (1140 square feet) and three-bedroom (1375 square feet). Entrances are in the middle on the ground floor, with a scissors stair in the corner—a centralized plan that gives immediate access to all parts of each floor.

The floor plans are tight, but laid out for maximum useability. Each unit has a dining room, living room and master bedroom, and windows allow natural light into every room. Despite the disciplined layout, there is plenty of storage space. Solar collectors are also planned to provide solar-heated domestic hot water. The water would be stored in oversized hot water tanks. The solar roof monitors shown on the site plan and elevations would be simple framed boxes with windows in north and south walls—somewhat like vertical skylights. The monitors would provide light and solar heating to northern areas of the units and function as natural ventilation "chimneys."

Heating and Cooling

Mogavero and Unruh think of their Sacramento climate as "moderate." The winters are rainy and overcast with temperatures averaging 40° F. to 50° F., making it important to collect diffuse solar radiation and install thick insulation. The summers are dry. The days are very hot (up to 110° F. with design temperature at 102° F); yet nights cool down to 60° F. It is nearly a desert climate, and these temperature extremes make house cooling as crucial as heating.

The main source of passive solar heating is the south wall of each unit, which will have double glazing and fixed overhangs for summer shading. The relatively shallow depths of the units (about 24 feet) allow excellent solar penetration. Solar heat will be stored as follows: in thermal mass comprised of a perimeter-insulated and quarry-tiled concrete slab; in 1⅛ inches of plaster over gypsum wallboard on ceilings of both levels. The plaster will be gypsum with extra sand and will weigh about 120 pounds per cubic foot. Total mass surface in each unit is approximately 2,600 square feet, and the thermal mass is used for both heating and cooling. A simple duct with a thermostatically controlled fan will redistribute the stratified excess solar gain in the winter.

One aspect of the thermal mass unique to the Windcreek project is that it is spread out. The architects reasoned that this surface mass would absorb and radiate heat or night coolness better in the Sacramento climate; the cool nights would tend to make only the very outer layer of the thermal mass effective. Besides using the same thermal mass for cooling, each unit will have a large-volume, whole-house fan to assure the required air changes per hour for summer night cooling.

The passive thermal mass technique compares favorably with mechanical air conditioning from a cost point of view, even though electricity in Sacramento costs only 2½ cents a kilowatt.

Construction

The Windcreek foundations and ground floors are economical concrete slabs, surfaced with quarry tile. This economy is matched by two-story construction, which provides twice the square footage of living space for the same foundation, roof framing and finish costs. Single-piece, precut truss joists, supported by an intermediate bearing wall, are another economical construction feature.

Because the south walls are mainly transparent, intermediate bearing walls are needed to meet seismic requirements and to help solve lateral stress problems. With the intermediate walls, the costs of extra hardware and plywood for lateral stiffening are reduced. The roof framing has a ½-inch-per-foot roof pitch while maintaining an integral flat ceiling. With short spans, the header and beams can all be a standard 4- × 12-inch size.

Three Heating and Cooling Options

Because the Windcreek project is unbuilt, the developer is considering three heating/cooling options, each of which can be provided within the same architectural scheme. Energy consultants have analysed both the thermal and economic consequences of each option, but a final choice has not been made. (In making their analysis, the consultants used California calculation procedures; these are standardized methods of measuring the energy budgets of construction projects.) The three options are:

Full passive solar heating and natural cooling. The south walls would be designed to have all the glass needed for passive solar heating. Heat would be stored in the thermal mass. Cooling would be accomplished by night ventilation across the mass using the whole house fan. Two gas wall heaters in each unit would serve as backup. This option requires south glass to be 18 percent of total floor area. During summer nights the house fan would provide 15 air changes per hour.

Moderate passive solar heating and natural cooling. Some moderate use of solar heating and summer night ventilation cooling as described in option one would be combined with a conventional heat pump. This would be the most expensive option, but may be most likely to appeal to buyers because of familiarity and ease of control. South glass would be 14 percent of floor area. During summer nights, the cooling of the interior would require 12 air changes per hour.

A combination of options one and two. The interior thermal mass would control the cooling, and reduced south-wall glazing could be installed to match the capacity of the mass for heating. Backup heat could be provided by a conventional gas furnace with heating ducts. Glass area would be 11 percent of floor area thus reducing incidental unwanted heat gain in summer. Summer cooling is based on night ventilation of six air changes per hour.

The additional costs of the electricity for the heat pump and the gas for the gas furnace make the first option (full passive) the most economical. Its major disadvantage for potential buyers is the somewhat greater fluctuation of house temperatures and the lack of instant temperature control, characteristics common to passive solar homes. However, the savings on the electric bill (about $250 annually) compared to a standard new structure may still make this option the most desirable to many.

Sun-Up— The Solar Sufficient Dormitory

BUILDING

Sun-Up Dormitory
Rocky Mountain School
1493 Road #106
Carbondale, Colorado 81623

DESIGNER

David Finholm & Associates, architects
David Finholm, principal architect
Tom Crews, job captain
Box 2839
Aspen, Colorado 81611

CONSULTANTS

Ron Shore
Doug Davis
Thermal Technology Corp.
Snowmass, Colorado 81654

BUILDER

Diemoz Construction Company
318—20th Street
Glenwood Springs, Colorado 81601

STATUS

Occupied 1979

Estimated Building Energy Performance

BUILDING DATA

Heating degree days	7,340 DD/yr.
Floor area	4,461 sq. ft.
Volume of conditioned space	45,120 cubic ft.
Building cost	$231,900
Cost of special energy features	$42,000
Total building UA (day)	998 Btu/°F./hr.
Total building UA (night)	563 Btu/°F./hr.
Heating	
Area of solar-heat collection	1,349 sq. ft.
Solar heat	90 percent
Auxiliary (wood)	2,717 Btu/ sq. ft./yr.
Hot Water	
Need	57,600 gallons at 140° F. (incoming 40° F.)

ANNUAL ENERGY SUMMARY

Nature's Contribution	
Heating	63.0×10^6 Btu
Hot water	45.1×10^6 Btu
Electrical	5.8×10^6 Btu
Auxiliary Supplied	
Electricity and firewood	42.0×10^6 Btu/yr. (or 9,408 Btu/ sq. ft.)

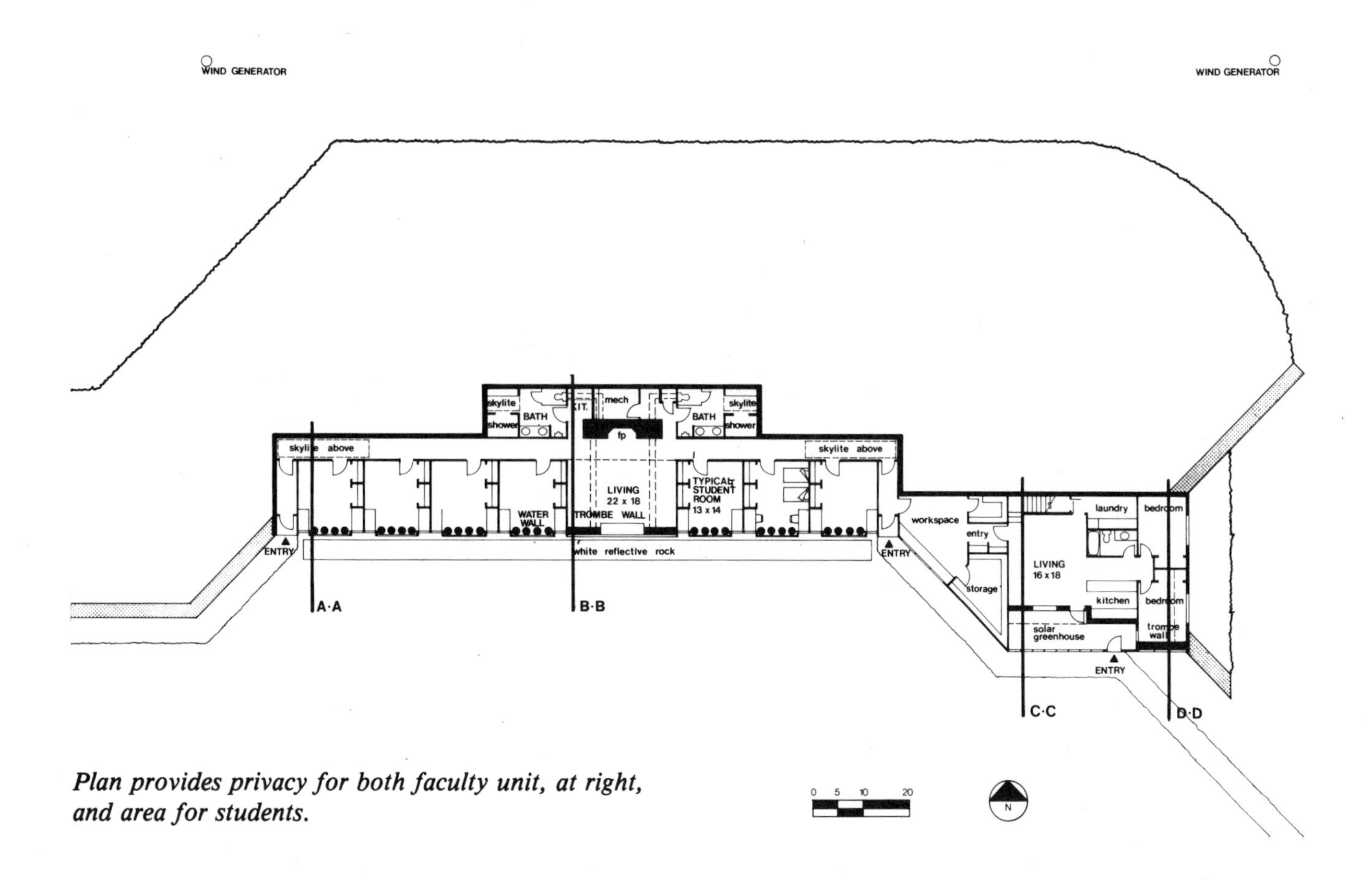

Plan provides privacy for both faculty unit, at right, and area for students.

The Colorado Rocky Mountain School is an exemplary boarding school in many ways. This exceptional solar-sufficient dormitory designed as a living educational tool began as a student project in a class taught by Ron Shore on energy efficiency and physics. As the hypothetical academic exercise progressed, it was discovered that grants were available to help fund the construction of such a project. Massive group efforts were started to accomplish all the requirements of grants in the time available. From this, an extraordinary building resulted. This snug earth-embraced 5,600-square-foot residence is much more than a warm and rugged shelter in a harsh Rocky Mountain climate.

A Textbook of Passive Design

The purpose of the dormitory was to be an educational tool to demonstrate all the passive concepts that were currently being used in buildings. These include Trombe walls, water walls, direct and indirect gains, passive solar greenhouses, and a passive hot water collector, which should be called a non-imaging specular reflector collector. The movable insulation, which is a self-inflating reflective operable curtain, is powered by direct current electricity from photovoltaic cells. By using a carefully designed circuit network, the amount of electricity required is minimized when all the curtains are asking to come down at the same time because the building is getting too cold.

The rigor of a cold mountainous climate required equal rigor in design and construction. The shape of the building was carefully designed. The building was bermed into the ground on three sides for economy and energy-efficient reasons, using the earth as a natural insulation. The students themselves worked out the basic heat loss loads and gain calculations to determine which shape, how much glass, and the proper ratio of insulation

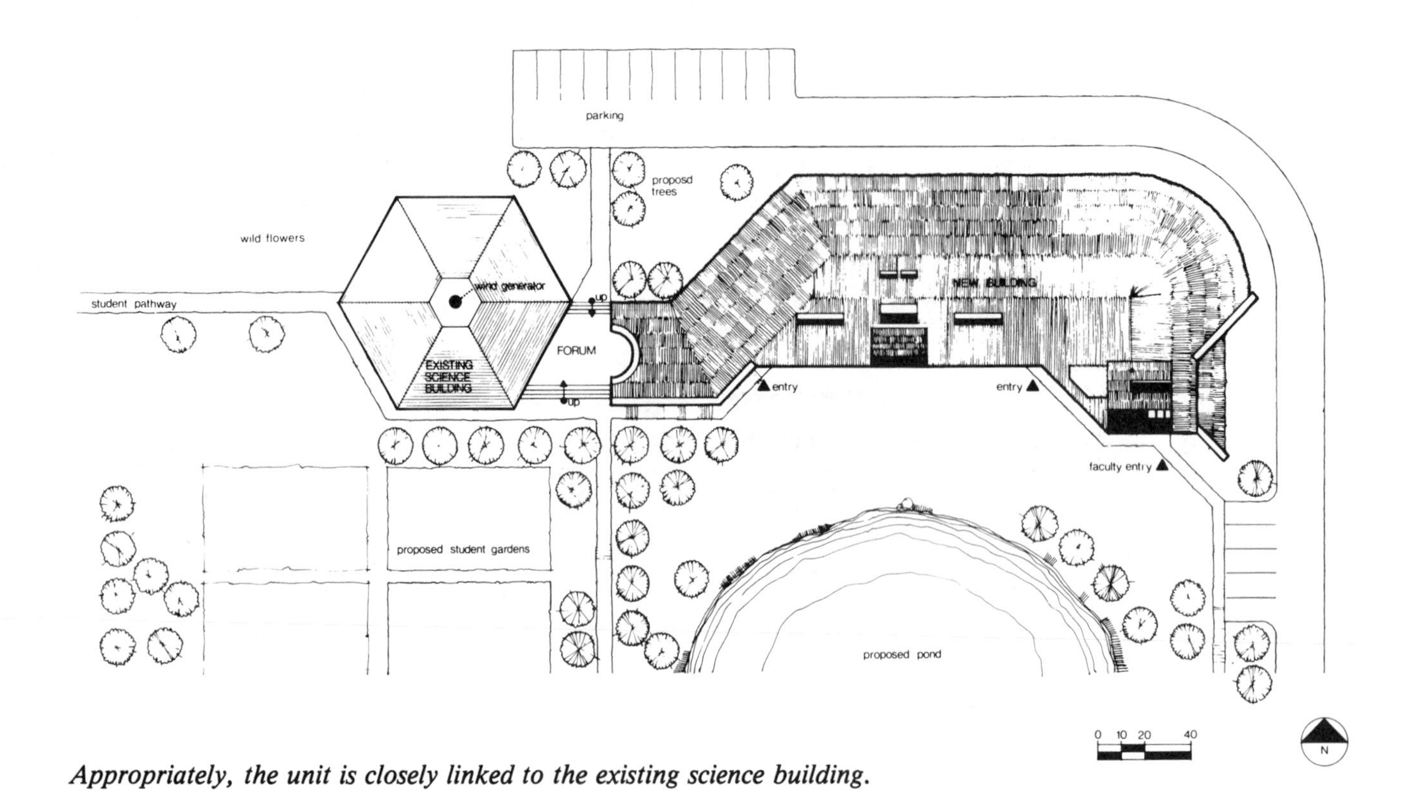

Appropriately, the unit is closely linked to the existing science building.

would produce a building that would use little or no back-up energy from fossil fuels. This was the initial purpose of the project as a study. But the study got built.

Education Through Construction

The project began when a couple of students, disgruntled over what they perceived as the shabbiness of dormitory life, built a small cave-like dwelling in the side of a hill, found a wood-burning stove, and moved in. With a dirt floor, minimal light, and no electricity or running water, they felt it was still preferable to their dormitories.

When the need for more student housing at CRMS became critical, the cave became an inspiration. Ron Shore and his students decided that a good class project would be the design of a student dormitory that was energy self-sufficient. The initial design housed only four students. But when the administration began to show an interest in the project, the plans became more ambitious, expanding to house fourteen students, with a faculty residence attached.

A student who enrolls in the Colorado Rocky Mountain School can expect to be exposed to heavy doses of experiential learning in addition to a demanding academic schedule. Much of the maintenance around campus is done by students who are assigned work projects, and outdoor education classes are given throughout the school year. In March, the school effectively closes down, while students go off on individual projects, ranging from painting a mural in a prison to extended forays into the wilderness.

In earlier years Ron Shore had involved students in weatherization projects such as caulking, insulation, storm windows, double doors, and weatherstripping on school buildings. One class built a solar hot water heater for the pottery building, and another put a hot air collector made from water-filled aluminum cans in the music listening hut.

In 1976 they designed and built a passive solar greenhouse which produces a year-round supply of fresh vegetables for the school.

But the solar dormitory project involved education well beyond the physics classroom. City, state, and county bureaucracies had to be convinced that the dormitory was feasible. A permit was needed from the building department; electrical inspectors, horrified by the idea of a room full of huge batteries for storing electricity produced by wind generators, had to be taught the inner workings of the direct current system; and the local utility company had to give a permit for the wind generators. Ron and a couple of students spent a winter vacation working on the endless pages of cost projections necessary for obtaining grants and donations.

Students in Design and Construction

While the bureaucrats slowly digested these seemingly revolutionary concepts, students were designing and fact finding. The first step of the design phase was choosing from among six possible sites for the building. Each site was analyzed for a variety of factors such as: Is there enough sun and wind? Is the ground already productive? Would the students living there feel isolated?

Local architect David Finholm, who worked on the building through completion, taught the students how to put their design ideas down on paper, and then to create cardboard models. Some of the designs were downright spartan, with cramped spaces and no creature comforts. Others included indoor swimming pools and hot tubs, exercise rooms and tennis courts. The model that came closest to the real world of time and budgets, without sacrifices in lifestyle, was a one-story rectangular structure with south-facing windows and earth on the north-facing walls and roof.

Facts and figures came next. Again, the right questions had to come first. What are the energy needs of fourteen students and a faculty family? What local energy sources are available? The students had to figure out, down to dollars, degrees, and decimal points, what their energy needs were. They handed out questionnaires to other students and put charts in dormitory bathrooms to determine how many times a day the toilet was flushed and how often and for how long sinks and showers were used. They went into darkrooms and plugged in different kinds of lightbulbs until there was enough light for reading and writing by beds and desks. The passive solar elements of the building were tested with small thermal models. A wind power expert was consulted on the most efficient wind towers and generators available.

As the students worked on the design of the dormitory, they began to apply the reading they had been assigned—among others, *Soft Energy Paths* by Amory Lovins and *The Closing Circle* by Barry Commoner. The key concept was the use of appropriate technology.

Celebrating the Soft Energy Path

The appropriate "soft energy path," as they learned, was to decentralize energy consumption and put it back into the control of the individual by utilizing the immediate environment to adapt to the climate. For instance, there was an abundance of wood available; therefore a wood-burning furnace for heating hot water was planned. Snow was used to advantage as an insulation on the earth roof and as a reflective surface outside the south-facing windows. In contrast, in Detroit, the appropriate resource may have been used oil; in the Northeast, hydropower; in West Virginia, coal, and on the coasts, wind.

Another myth shattered was that using renewable energy sources efficiently means going back to caves and candles, or living in an ugly or uncomfortable building.

It didn't take the students long to realize that the less energy used, the easier it is to supply. They set up systems that students could maintain themselves, passive systems with no moving parts, and systems that would be architecturally easy to change or adapt in the future.

Questionnaires made it clear that one feature sorely needed in the other dormitories was common space. In Sun-Up, this was planned as an open, spacious room with an energy-efficient fireplace and a spectacular view of Mount Sopris. Bathrooms, a utility room, and a small kitchen were placed behind the common room, against the north-facing wall. Everyone agreed that students and faculty should be separated from each other, so a battery storage room and a workroom, designed to accommodate bicycles and cross-country skis, were put in between the two areas.

The simplicity of passive solar heating is celebrated by Sun-Up. The building, like a flower, opens up to sunshine during the daylight hours and folds up at night. Amory Lovins claims that any home located outside the Arctic Circle shouldn't need anything but passive solar heat if it's built properly. Sun-Up confirms this theory. The handsome dormitory is so well constructed that the rooms tend to get stuffy because there are no cracks to make air circulate. If anything, the problem has been overheating, and the baseboard systems intended for backup heating are rarely used in winter.

The dormitory consists of seven student rooms with two students each, a living room shared by all students, two bathrooms, and a small space for cooking tea and soups for lunch. Adjoining this is a 1500-square-foot apartment used by the faculty monitor. This includes three bedrooms, two baths, and living, dining, and kitchen spaces. Adjacent is the solar greenhouse by which it obtains most of its heat.

A Catalogue of Energy Design Techniques

Sun-Up as a solar dormitory uses not one but several passive techniques to meet its space heating and domestic hot water needs. It is earth-bermed on the north, east, and west sides, and the roof is partially covered with sod. There are two main living areas separated by a vestibule entry. The south wall of the student living area consists almost entirely of south-facing glass, and is a mixture of direct gain, water walls, and a Trombe wall. Additional interior mass is provided by the concrete floor and north, west, east, and interior walls. More solar gains are brought into the building through south-facing clerestories. The north walls of these clerestories are covered with reflective tile which directs the solar radiation onto the interior mass wall. A large greenhouse with a concrete north wall located on the south side of the faculty living area acts both as a buffer zone against heat loss and as a solar collection area. Direct gain windows comprise most of the remaining area of the south wall of the faculty residence.

The structure of glued, laminated timber post-and-beam is attached to the 12-inch reinforced concrete back wall that serves as a retaining wall for the earth-banked north side. A plywood roof deck, insulated and waterproofed with a continuous membrane, supports the 18-inch ground and grass sod. The high curved hoods of the north solar scoops have a mirror-like lining of specular aluminum foil to reflect solar energy in concentrated beams to a 2-foot strip of mirrors. From here the sun's energy is spread over the thermal mass of the solid masonry north wall. Thus the sun shines into the north corners of the dormitory through the earth cover, as well as into the opening of the open funnel of south wall windows. The exposed dark colored concrete floor slab is on a gravel bed with 2 inches of foam underneath to complete the insulation wrapping of the entire structure.

All large glazing areas are outfitted with *Insulating Curtain Wall,* an automatically operated insulating curtain. Power for the curtains is supplied by two photovoltaic arrays located nearby. Photovoltaic power is also used to run three ceiling fans which move warm air from where it collects at the ceiling down to the storage mass.

Two 12-foot-diameter wind generators supply 32 volt dc power which is used for the dorm's lighting system. Research into wind velocity, frequency, and direction around the building site indicated that using the wind to generate electricity was not satisfactory in terms of cost. Nevertheless, wind-generated electricity was installed because all involved agreed it was important to learn how to tap and how to live with this important resource. Conversion from direct current to alternate current uses substantial amounts of energy, so the building was designed for direct current, with the exception of a few outlets in the faculty quarters.

Electric clocks, hair dryers, and individual radio and stereo systems were deemed extraneous.

Little is seen of the building in this view from the northwest.

Since student surveys had shown that they would rather do without light than without music, a central stereo system located in the common room provides music through speakers and channels wired into the rooms.

Each room has a circuit breaker that will blow a fuse if every light in the room is turned on. This way the whole dormitory isn't penalized if one room is using more than its share of energy, and every student becomes aware of what he is consuming.

The Energy Economics of Self Sufficiency

One concept of the building design was to reduce energy. From this premise, the hot water system can be explained. As the cold water supply at 42 ° F. enters the building it is stored in the mass of the building in a tank next to the furnace to raise its heat naturally from the building. Then it is heated through copper tubing coils under the wooden slats on the floor of the showers. By the time it is heated in the solar hot water collector on the roof, the energy requirement has been reduced considerably just by thoughtful routing of the cold water.

As the waste water is drained from the building, from the showers and the sinks, it is stored one more time to be used to generate heat within the building and, secondly, to be used to flush the toilets, operated by a simple hand pump. What's not used by the toilets can be used to water greenhouse plants and the student gardens.

This type of thinking made the building very successful in terms of minimum energy demands from outside sources. From the time the building was occupied in October, 1979, until its dedication in April, 1980, the outside energy requirement was only $14 for electricity. Firewood from the school yard makes up any heating deficit.

During the winter of 79–80 the building used two cords of wood to heat auxiliary spaces and domestic hot water. Assuming 19 million Btus per cord of pine and a 65 percent efficiency for the stoves, 25 million Btus of auxiliary heat were used in both categories in the first winter season. Observations by the students show that the Rumford fireplace was used mainly for aesthetics, the back-up hydronic system was used only once when the insulating curtains were inadvertently left up during Christmas vacation, the airtight stove in the faculty section was used occasionally, and the wood-fired back-up domestic hot water heater was fired up every few days during cloudy weather.

School Solar Dormitory

BUILDING

Fountain Valley School Solar Dormitory
Colorado Springs, Colorado

DESIGNER

Alfred von Bachmayr
1019 East Moorehead Circle
Boulder, Colorado 80303

BUILDER

Contractors, Managers, and Consultants, Inc.
1620 Market Street
Denver, Colorado 80202

STATUS

Completed May 1981

CONSULTANTS

Steve Baer, solar
Zomeworks Corp.
1221 Edith Northwest
Albuquerque, New Mexico 87102

Don Felts and Associates, mechanical, electrical
1221 Edith Northwest
Albuquerque, New Mexico 87102

John Wulfmeyer, architect
2345 Seventh Street
Denver, Colorado 80211

Mark Upshaw, architect
2580 West College Avenue
Denver, Colorado 80219

Estimated Building Energy Performance

BUILDING DATA

Heating degree days	6,300 DD/yr.
Area of conditioned space	10,070 sq. ft.
Volume of conditioned space	81,600 cubic ft.
Building cost	$678,000
Cost of special energy features	$52,000
Total building UA	1,622 Btu/°F/hr.
Heating	
Area of solar-heat collection	1,110 sq. ft.
Solar heat	75 percent
Auxiliary heat, natural gas	37,160 Btu/sq.ft./yr.
Cooling	
Natural ventilation, movable reflector	
Auxiliary ventilation (4 horsepower fan motors)	(1,220 Btu/ sq.ft./yr.
Cost of auxiliary	1¢/sq.ft./yr.
Hot Water	
Need 360,000 gallons at 110°F. (incoming 50°F.)	
Shower water heat recovery	25 × 10^6 Btu/yr.
Auxiliary, natural gas	180.14 × 10^6 Btu/yr.
Cost of auxiliary	$653/yr.
Lighting	
Need four hours/day × 240 days × 1.4 Kw/sq. ft.	
Thermal equivalent of electricity used	4,587 Btu/ sq.ft./yr.
Cost of electricity for lighting	3.8¢ sq.ft./yr.
HVAC Fans	
Need 1.92 Kwh/sq.ft./yr.	
Thermal equivalent of electricity used	6,565 Btu/ sq.ft./yr.

ANNUAL ENERGY SUMMARY

Nature's Contribution	
Heating (solar)	290.3 × 10^6 Btu
Cooling	Not available
Hot water	91.2 × 10^6 Btu
Auxiliary Purchased	
Electricity and gas	250 × 10^6 Btu or 24,870 Btu/ sq.ft./yr.

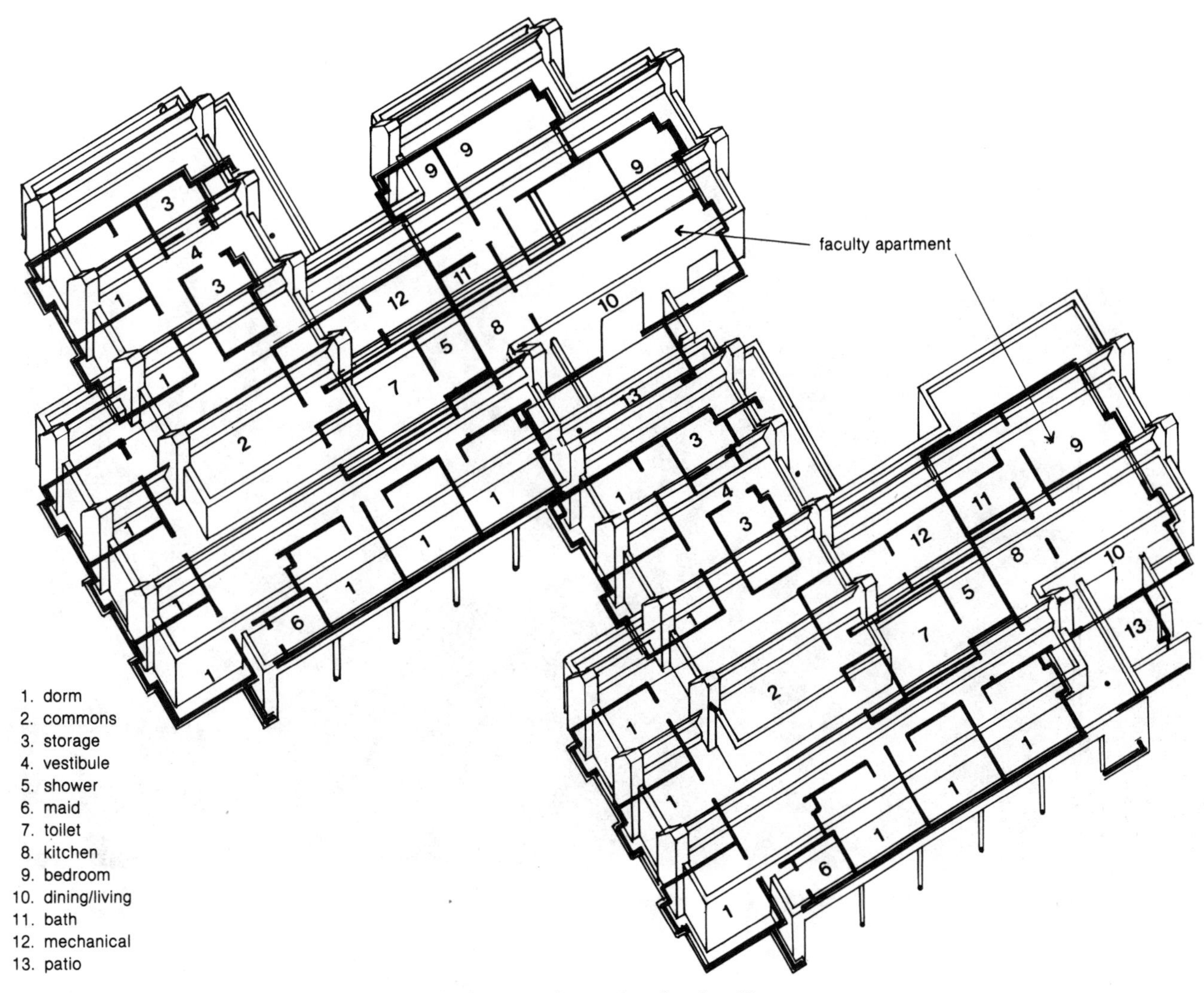

This plan provides space for thirty-two students and two faculty families.

Solar Dormitory for the Fountain Valley School

This dormitory for a private school houses thirty-two students and two faculty families. Its design is based upon a unique skylight concept. Passive solar space heating, solar hot water heating, and daylighting are all accomplished through the use of skylights and solar concentrating reflectors. Yet the style of the one-story, 10,000-square-foot building, completed in 1981 at a cost of $678,000, is in keeping with the traditional southwest architectural character of the Colorado Springs campus.

The Innovative Multi-Purpose Skylight

The architectural design is based on the performance of an innovative multi-purpose skylight.

During the colder months of the year, sunlight strikes a 4-foot by 14-foot skylight module and is beamed by a parabolic mirror through a 2-foot-wide aperture in the ceiling of each room. The

These units beam sunlight into the interior of the building.

aperture, covered with double-paned, iron-free glass, acts as a diffuser for the concentrated light coming from the reflective mirror. After passing through the diffusing glass, the sunlight strikes hot water pipes, on which are mounted blackened aluminum plates in parts of the building. By preheating, the energy demand on the building's hot water system is reduced. The diffused sunlight then enters the room below, illuminating the walls and floor. Because the walls are made of grout-filled concrete block, the solar heat is retained in thermal storage. At night, the floor, walls, and ceiling gradually release the stored heat to the living areas. Unit heaters provide back-up heat and destratify the air.

During the summer, the mirrors in the rooftop collectors are tilted down to reflect sunlight away and to prevent solar heat from entering the building. Ventilator turbines in the roof help cool the roof at night, and an exhaust fan in each commons room draws night air into each room through open windows, cooling the interior masonry. During the day, the masonry absorbs heat from the living space, keeping the dorm cool.

The innovative skylight is a fresh solution to typical solar design problems. With it, a high solar fraction and desirable natural daylighting are achieved without the use of large expanses of glass or movable night-time insulation. Moreover, by providing heating and lighting from overhead in such a one-story building, there is design flexibility in plan arrangement, a concept that has applicability to many other building types, both residential and commercial, both new and retrofit.

Other energy-saving features have been incorporated. Wood-burning stoves in faculty apartments and in the commons areas can be used to provide backup heating. All windows are double-paned and have thermal-break aluminum frames for energy efficiency and low maintenance.

In addition to using the natural energy resources of the site, the dorm also was designed to reflect the social needs of the campus. Two faculty apartments are adjacent to the dormitory and configured to minimize noise transfer from one room to the next. No two windows of student rooms or faculty apartments face each other, creating a sense of privacy.

Estimating the Energy Budget

A preliminary energy-cost analysis by consultants Steve Baer and Don Felts confirmed that this unique solution would be financially feasible and should be built. They estimated that the well-insulated building envelope would have a heat loss on the order of 40,000 Btu/square foot a year. Of this amount, the passive solar heating would provide as much as 30,000 Btu/square foot a year. The remaining 10,000 Btu/square foot a year would be made up by a conventional heating system.

For comparison, a typical building of this type in the Colorado Springs climate would require on the order of 60,000 Btu/square foot a year for space heating, and all of the energy for the space heating, water heating, and lighting would have to be provided in a conventional manner.

In addition to the space heating, the passive solar skylights would virtually eliminate the need for artificial lighting during the daylight hours, an annual saving on the order of 6 Kwh per square foot a year in lighting energy.

The domestic water heating energy requirement for the entire building was estimated to be as much as 180 million Btu per year. The solar skylight water pre-heater would provide as much as 100 million Btu per year of this energy requirement.

The consultants' research indicated that in 1980 Fountain Valley School paid $4.35 per net usable million Btu of heat, and 2.9 cents per Kwh of electricity. These rates were favorably low when compared to other cities in the Southwest. At these rates, the energy cost savings of the proposed building as compared to a conventional building would amount to as much as $4,400 per year. If the cost of energy inflated at the rate of at least 15 percent per year for the next five years, the energy savings would amount to approximately $8,900 per year at the end of the fifth year.

A Solar Plan That Meets Many Needs

The innovative skylights allow the plan of interior spaces to respond to other needs than thermal. Thus dormitory rooms can face west as well as south. Each set of eight dormitory rooms is arranged around a common room. On the east side is a faculty apartment which is completely self-contained with its own separated courtyard patio. Faculty families have complete privacy while maintaining supervisory capabilities.

The arrangement allows each cluster of accommodations and even each room to develop its own character within the regularity of the whole building. The dormitory building itself is designed for future expansion, as well as to respect the existing campus plan. It provides closure for the existing quadrangle and still allows drainage for occasional heavy surface runoff. Set between mature trees, the dormitory floor levels step into the existing land contours and maintain a low profile in keeping with the existing buildings surrounding the quadrangle.

Shaping the Design Parameters

In determining the program of a dormitory for the students at Fountain Valley School, several design parameters were considered. Aside from aesthetic and site-planning considerations, the energy use of the building became the major determinant. An energy-saving design could not only reduce operating costs, but could also permit the building to serve as a model for students who want to learn about the physics of energy. With the cost of fuels

Sunlight streams through glass at bottom of photo.

increasing, it was felt that all the residents in the new dorm should have the opportunity to learn about energy use since it would affect them in coming years.

The dormitory was designed to be responsive to climatic conditions as a means of saving energy, in contrast to one that is a sealed environment and operates at a constant temperature due to major inputs of fossil fuels. The temperatures in this passive dormitory gently fluctuate daily in response to the conditions outside. For example, in the winter, the building will tend to be warmer in the afternoon and early evening than in the morning. In periods without sun, the building gradually cools to a temperature at which the back-up heat comes on. For the occupants of the building, an additional sweater may be necessary after a cold night or a period without sun.

The new dormitory is meant to teach the students how to adapt to a living environment which is not always automatically within the ideal comfort zones. Once they learn to adapt, their future homes may use less energy due to their learned ability to be flexible and responsive to their environment.

Thermal Efficiency in Construction

Thermal efficiency is also enhanced by partially burying the building: the north side has no windows and has an earth berm 4 feet higher than the floor level. All walls are 8-inch reinforced load-bearing masonry block with cavities grouted solid for thermal storage. On the exterior, the walls are covered with 3-inch rigid styrofoam insulation with a stucco finish to maintain the Southwest character. The roof structure consists of wood truss joists spanning 28 feet. A composite roofing membrane over tapered rigid insulation provides water proofing between the skylight modules. Within the joist cavity there is 12-inch fiberglass batt insulation (R-38) so that the roof construction has a thermal rating of about R-40.

Within the dormitory areas, an exposed hard trowel concrete slab has integral color and an indented grid for a tile-like appearance. In the faculty apartments, the floors are covered with red quarry tile. Both flooring materials provide important expanses of thermal storage.

Realizing an Educational Asset

This design provides a stimulating example of traditional Southwest architectural character relieved by a sparkling skyline of mirrors that marks the presence of solar innovation. The construction of the Fountain Valley Solar Dormitory was by conventional contractors. Although an exemplary educational facility in its combination of energy and learning goals, the building was realized within a modest budget of money and time. Already, the facility has settled into the land and has become an integral part of the school campus and curriculum.

Manhattan Loft Conversion

BUILDING

Manhattan loft conversion

OWNER

National Railway Publication Company

DESIGNER

Banwell White & Arnold, Inc., architects
2 West Wheelock Street
Hanover, New Hampshire 03755

STATUS

New York City variance denied, project halted

CONTRIBUTORS

David P. Helpern Associates, associate architects
9 E. 41st. Street, New York, New York 10017
Ambrosino & DePinto, P.E., mechanical and electrical engineers
225 W. 34th Street, Suite 2107, New York, New York 10001
Alvin O. Converse, solar engineer
Thayer School of Engineering, Dartmouth College, Hanover, New Hampshire 03755
Robert O. Smith & Associates, P.E., computer simulations
55 Chester Street, Newton, Massachusetts 02161
G.K. Associates, computer simulations
157 Stanton Avenue, Auburndale, Massachusetts 02166

Estimated Building Energy Performance

BUILDING DATA

Heating degree days	4,848 DD/yr.
Heating degree hours	104,000 °hr./ Base 61.5/68°
Cooling degree hours	9,159 °hr./Base 71.5/78°
Building area	128,000 sq. ft.
Building cost	Not available
Total Building UA	12,017 Btu/ °F./hr.
Heating	
Area of solar-heat collection glazing	
Solar heat	23 percent
Auxiliary heat (oil)	13,950 Btu/ sq. ft./yr.
Cost of auxiliary heat (at $1 per gallon)	10¢/sq. ft./yr.
Cooling	
Natural ventilation	
Auxiliary cooling (oil)	7,295 Btu/ sq.ft.
Cost of auxiliary cooling (at $1 per gallon)	5¢/sq. ft./yr.

Hot Water

Need 2,892,000 gallons at 120° F. (incoming 50° F.)

Solar heat	30 percent
Auxiliary fuel (#6 oil)	2,149.2 × 10⁶ Btu/yr.
Cost (at $1 per gallon)	$14,820

ANNUAL ENERGY SUMMARY

Nature's Contribution	
Heating	533×10^6 Btu
Cooling	Not available
Hot water	934.3×10^6 Btu
Auxiliary Purchased	
Oil	$7,053.2 \times 10^6$ Btu or 55,100 Btu/sq.ft.

Ten units such as these were planned for floors 2-12.

It is relatively easy to design a high-performance, passively heated new house for a suburban or rural lot. But it is usually assumed that urban applications are difficult even for new buildings and that our existing cities are impossible for passive retrofits. The conversion proposal in New York City of a 13-story industrial loft into 116 passive solar apartments is a convincing prototype of both design strategies and implementation issues.

Among the central questions are whether existing under-utilized, high mass buildings can be converted successfully to energy-efficient residential units at a substantial net energy savings by the customary demolition and new construction process. In addition, the inherent efficiency of contiguous multi-floor housing units and the economies of urban transportation suggest an increasingly important concept of total energy efficiency in the potentials of urban living.

These questions were addressed comprehensively by both architects and owners. Unfortunately, the project was shelved because of denial of a city zoning variance. Nevertheless, the project had sufficient significance that it was further studied in depth by the Brookhaven National Laboratory and published as a U.S. Department of Energy *Case Study*. The findings reinforced the design conclusions and investment advantages of the private study. They stated that the "renovated building is seen to have an infinite energy advantage over the demolition/replacement building."

Massive construction is excellent for thermal mass and acoustical isolation.

A Microcosm of the Conversion of Industrial Lofts

The existing industrial loft at 424 West 33rd Street is not unlike many industrial buildings in the nation's older inner cities that were built in the first half of the 20th century. Distinguishing characteristics are massive steel and concrete structural systems for high industrial floor loadings, thick masonry walls, relatively large glass areas, high ceilings, and spartan architectural design.

The situation at 424 West 33rd is a microcosm of the industrial loft problem in Manhattan. Partially rented with several vacant floors, the building was unable to generate adequate income for years. The owner, eager to convert a long-standing liability to an asset, decided to investigate the feasibility of a residential conversion under the J-51 program. That 1955 New York tax abatement program was designed to encourage owners of existing residential properties to upgrade their buildings. It was later expanded to include conversions of industrial lofts to residences.

The Site

Because of an opening in the city-scape over the Amtrak yards just west of Penn Station and the Post Office, the south side of the building is open to the sunlight. In Manhattan the street grid is 28° 30′ west of true south. The building is 130 × 90 feet, with the long side facing south, and is 160 feet high. Shading doesn't begin until after 3 p.m. in December from the building neighboring to the west.

Twelve stories and a penthouse, or 20,000

square feet of potential solar collector area, are offered by the south wall. This plus an additional 11,000 square feet on the roof provided a strong impetus for a solar-assisted solution.

Advantages of the Structure

The massive 1912 structure is a concrete-encased steel frame with 9-inch reinforced concrete slabs and masonry walls 20 inches thick at the base and 12 inches on typical floors. The floor is designed for a load of 240 pounds per square foot, far in excess of standard requirements for residences but excellent for acoustical isolation and thermal mass.

Because of the building depth and floor-to-floor height, such coveted amenities as ample storage, walk-in closets, large living rooms and bedrooms, 12-foot ceilings, and the general impression of space can be provided. This is of great value to the developer who can spend less in construction and charge rents roughly equal to new apartments, depending on the location of the building.

This building will never become an entry in the National Trust's list of historic buildings despite its ornamented pilasters at the 12th floor. Its lack of historic or artistic significance permits the greatest flexibility for adaptive reuse and energy-oriented retrofitting. An analysis of embodied energy shows the retrofitted building has an infinite energy advantage over a hypothetical replacement building designed to meet current energy codes.

Thermal Performance

A continuous indirect gain sunspace was the preferred solar design for this building. Sunspaces are 6 feet wide and run the length of each south-facing apartment. The proposed outer wall is a combination of fixed and operable glass, floor to ceiling. The inner wall of the proposed sunspace is approximately 50 percent masonry for thermal storage and 50 percent glass for view and daylighting. During the heating season, with the outer window wall vents closed, the inner wall is never exposed to outside wind conditions, greatly reducing infiltration and conduction losses. Since the winter sunspace temperature is always above outside temperature, the number of degree-days for heating is dramatically reduced by passive solar gain.

The sunspace transfers energy into the apartment in the winter by direct gain and by radiation from the masonry wall. A primary advantage of a passive sunspace design is its performance on days with only diffuse sun or medium overcast skies. Under these conditions, there will be little transfer of energy to the apartment, but the sunspace will warm up sufficiently to virtually eliminate heat loss from the apartment. Auxiliary heating of the south-facing apartments in January is predicted to be only during late night and early morning hours.

The sunspace provides a buffer that minimizes the role of the occupants while deriving maximum fuel savings. They are only expected to keep exterior sunspace windows closed during the heating season and open during the cooling season. Night insulation is an individual option that the energy-conscious tenant may wish to exercise; the performance predictions do not assume the use of any thermal draperies. In contrast, a direct gain space is more likely to suffer from the intervention of the occupant, who will understandably draw blinds to reject glare and overheating from the low winter sun. Further, in a direct gain space, the building mass is likely to be negated by carpet and furniture, defeating the inherent benefit of the south-facing apartment. Summer sun will enter the conditioned space of a direct gain apartment, adding significantly to the cooling load, but the sunspace ceiling provides an overhang that intercepts the high summer sun and prevents it from getting into the apartment proper.

Cost and Benefits of Sunspace

The sunspace is created by removing the existing south windows and masonry infill, exposing the existing structural frame, reglazing the openings with new double-glazed operable sash, and constructing a new inner south wall 6 feet inside.

The sunspace floor is dark tile over the existing 9-inch concrete slab. The proposed new walls separating the sunspace from the apartment are to be 8-inch concrete blocks with the cores filled solidly with mortar. The block is dark and split-ribbed, providing increased surface area exposed to the sun, as well as an attractive interior texture.

The sunspace provides amenities over and above energy efficiency, such as year-round gardening opportunities and a climate similar to a Florida winter. The real estate consultant believed that the sunspace was a definite marketing asset, adding rental value to the apartments. In a cost benefit analysis the sunspace does better than a 100 percent direct gain scheme wherein south wall demolition is also required and heat losses are high. The sunspace does less well relative to a conventional scheme with south wall demolition and very low south wall losses. But the added value of the aesthetic amenities of the sunspace essentially enables it to pay for itself. The sunspace apartment requires only 35 percent of the heating load of the conventional south-facing apartment.

Solving the North Wall

There was some concern that the north-facing apartments would need special features to compensate for being deprived of the sun. This prompted some brief consideration of a north-side buffer space similar to the sunspace. The idea was abandoned as too costly, although it would be an energy-conserving feature to some degree. A continuing concern of the architects and engineers has been the transfer of excess south-facing solar gain to north-facing apartments. Several schemes were considered and rejected based on cost practicality. New York City consultants advised that these schemes generally involved the inventing of new technology, and a New York City J-51 renovation was not the appropriate testing ground.

Since the building's north wall has much greater heat loss than the south wall—a simple phenomenon almost always ignored in urban building design—the existing north window area is reduced by 40 percent, and the deleted glass area is replaced by an insulated panel, the resulting window being 3 feet, 8 inches by 5 feet, still an acceptable opening, a more or less standard apartment window size. New windows are used throughout the building. They are double-glazed casements, with awnings and fixed combinations. The sash and frame are made of aluminum with a thermal break.

The existing building is uninsulated. Various exterior wall insulation strategies were considered. Specifications call for 3½-inch fiberglass batt insulation on the inside surface of exterior walls, covered by gypsum wallboard.

Analyses of Thermal Performance

The proposed sunspace relies on the new masonry wall between the sunspace and the apartment as well as the existing sunspace floor and ceiling to absorb and store the solar radiation. This absorption by the masonry surfaces both reduces the peak temperature during periods of high incoming solar radiation and elevates the low temperature during the night. This dynamic behavior is of great importance and, therefore, a detailed computer simulation was employed to predict the performance and evaluate the effect of design modifications and alternatives.

The entire building was not simulated simultaneously. Instead, representative north-south apartment pairs were simulated. Although the thermal coupling of the south- and north-facing apartments is usually very low, this simulation of pairs of apartments allowed the investigation of design modifications to increase the thermal coupling between the south and north sides. Furthermore, the use of apartment pairs facilitated the comparison of north- and south-facing apartments, and thus, the impact of the sun.

Some of the most significant results of the study are displayed in the accompanying figure. Here the monthly heating and cooling loads for three different fenestration models for the studio apartment are compared. The lowest heating load is associated with the sunspace design, which has a large amount of glazing on the south wall, but has a massive wall between that glazing and the living space. The massive wall intercepts most of the insolation and stores a large fraction of it. This tends to stabilize the living space temperature, avoiding the high temperatures that would occur in the absence of the wall during the day, and supplying energy during the night to overcome the losses through the glazing.

Building Mass

In order to test the effect of the massiveness of the building, the sunspace configuration was resimulated with all mass elements reduced to only 10 percent of the original value. This essentially tripled the heating requirement of the studio apartment and increased the cooling requirement by 23 times. The heating requirement increased from 23.7 to 60.3, and the cooling requirement from 2.5 to 57.5. The massive building benefits greatly in its reduced cooling requirements by the fact that the ventilation air is conditioned to 65° F. and it cools the building during the entire 24-hour period. The mass in the building cools off and heats up only very slowly during the day and, hence, does not very often exceed the cooling set point of 78° F. Thus, the mass of the building plays a very significant role in reducing the cooling requirements. Likewise, in a low mass building the heat requirements are increased because the temperature of the building increases rapidly during the day, increasing losses, and there is little stored energy to carry the building at night.

The west-facing apartment is a corner apartment, and hence it has the large west wall through which it can be heated or cooled. The results of the simulation indicate that the effect of insulating the outside rather than the inside of these walls is negligible. In this building, the massiveness of the floors is sufficient, and one does not need to incorporate the massiveness of the side walls.

Choosing a HVAC System

The design and analysis of the heating, ventilating, and air conditioning systems were performed with the following two basic goals:

- The system design must be compatible with and take advantage of energy-conserving potential of passive solar energy.
- The system must be cost-effective in terms of life-cycle cost as it reflects both first cost and operating cost. The design should also be practical and one that can be effectively operated by the building maintenance personnel.

Design heating and cooling loads were developed based on manual calculations using standard ASHRAE methods. The peak loads were used for sizing HVAC equipment, system design, and cost estimating. These peak loads for the purpose of design *did not* take into account the solar collection contribution of the sunspace since the equipment must be capable of supplying the heating needs in the absence of the sun.

Estimates of the actual heating and cooling energy consumption were determined by both manual and computer network calculations. For the purpose of the various HVAC systems considered, the computer simulation numbers were used. The manual consumption estimates were used as a check of the reasonableness of the computerized estimates. Tenant electric, public light and power, and domestic hot water energy consumption were based on empirical data.

System Selection

The fancoil system with a low pressure absorption chiller has the lowest life cycle cost. This system also has one of the lowest energy consumption levels. Much of the cost benefit is derived from the difference in fuel prices. However, the system's longevity and low maintenance costs have much to do with the system's low overall costs. In addition to operating costs savings, this system also has these distinct advantages:

- The system will provide quiet operation within the apartment with individual room thermostatic control.
- With proper maintenance the system will have a long operating life. (Boiler and absorption chillers can last 30 years or more with necessary tube replacement.)

- Central air conditioning is a feature of a "quality" building. This will add to the building's market and rental value.

There are two disadvantages to this system. It does not readily provide the ability to meter each apartment for its heating and air conditioning and, therefore, cannot provide any direct incentive for renters to conserve. Secondly, the system does not allow for heating and cooling to be used simultaneously in different apartments. Systems that provide this ability are the water-to-air heat pump and the air-to-air heat pump.

Alternative

The water-to-air pump not only allows simultaneous heating and cooling, but uses it as a means for transferring energy within the building. This aspect of the system's performance makes it the most compatible with the passive solar concept. The operating costs could be reduced by about $8,000 per year if a gas boiler were used in place of the electric boiler. However, much of this savings would probably be offset by a change in the electric rate from 6.3 to 8.7¢ Kwh since the system would no longer be "all-electric." In commercial buildings, where large interior areas require cooling most of the time and heating is needed on the perimeter, the water-to-air heat pump provides an attractive approach.

Active Solar Collectors

An active solar system was analyzed that would assist in space heating, provide hot water, or even help to heat the quantities of ventilation air required by New York regulations. As a result of computer runs analyzing both net thermal energy collected and the comparative costs, the following conclusions were reached:

- The water-heating load is as large or larger than the space-heating load.
- Collectors are much more efficient serving the water-heating load than serving the space-heating load. For the size of collectors considered (and the size was made compatible with the surface area of the building), the increased energy savings when serving both loads was only about 5 to 10 percent greater than when serving only the water-heating load.
- Vertical collectors are much less effective than collectors mounted on an angle of 0° to 60° to the horizontal. Unless there is an overwhelming cost advantage to mounting collectors on the vertical south side, they should be mounted on the flat roof.
- Horizontal collectors do nearly as well as tilted collectors; therefore an attempt should be made to develop a low-cost system that is compatible with flat roofs.
- For this 12-story building, the roof area available for solar collection is limited and is much more effectively used for hot water heating than space heating. Very little improvement is gained by having the collector serve both loads. This would be all the more true for taller buildings.
- With inexpensive natural gas then available in New York, solar heating was not cost-effective. But if electricity were used, the capital cost would be recovered by expected fuel savings, and active collectors would represent a good investment.

Protecting Solar Access

The American Bar Foundation has concluded that "the single most difficult legal issue raised by the use of solar energy systems (is) guaranteeing access to light for solar energy collectors." Perhaps the fear that giving such a guarantee of solar rights for this building would prevent or hamper development of the air rights over the nearby Amtrak yards underlay some of the opposition to the project.

As has been noted, "A passive solar system has only one moving part—the earth moving around the sun." Active solar systems employ mechanical means to aid in heat transfer but are not

Five new floors of solar penthouse apartments were planned.

substantially more complex. Thus, their major problem may be as simple as solar access.

In high-density, high-rise urban areas like Manhattan, where passive solar design requires full protection for south-facing walls rather than merely the protection of south-facing rooftops, solar development may be precluded.

The Value of This Project

Both the design expertise and the analytical tools are available and well-honed for such a large-scale project. Similarly, the technology and hardware are ready. In this case it was the political climate and the impact of zoning regulation which were decisive, not the weather or the solar applications field.

By its completeness this project is not just a microcosm of the energy design in New York conversions, but of the issues that must be addressed in any urban application, whether a new building or a retrofit. Existing city buildings represent an enormous energy asset. If developed, the value of energy saved in existing and proposed buildings in the next 20 years could equal twice the cost of providing them with conventional energy in the same period. But the leadership role of cities in this process is yet to be seen.

From Ugly Derelict To Solar Swan

BUILDING

Simmons Building
Davol Square—Phase Two
Point and Eddy Streets
Providence, Rhode Island

ARCHITECTS

Beckman, Blydenburgh & Associates/
Architects, Inc.
116 Chestnut Street
Providence, Rhode Island 02903

DEVELOPER

Providence River Properties
P.O. Box 186
Providence, Rhode Island 02901

STATUS

Occupied Fall 1983

ENGINEERING

Mechanical Engineers
R.L. Horridge & Associates
168 Lavan
Warwick, Rhode Island 02888

Structural Engineers
The Yoder Corporation
P.O. Box 1622
Providence, Rhode Island 02901

Electrical Engineers
Gaskell & Associates
1033 Narragansett Blvd.
Cranston, Rhode Island 02905

Estimated Building Energy Performance

BUILDING DATA

Heating degree days	6,000 DD/yr.
Floor area at ground	6,500 sq. ft.
Total floor area	30,000 sq. ft.
Volume of conditioned space	945,000 cubic ft.
Total cost of renovation	$1,300,000
Cost of special energy features	$500,000
Total building UA	5,000 Btu/°F./hr.
Heating	
Net area of solar collection glazing	8,000 sq. ft.
Auxiliary heat (gas) cost	15¢ sq. ft./yr.

ANNUAL ENERGY SUMMARY

Nature's Contribution	
Heating	400×10^6 Btu
Auxiliary Purchased	
Gas & electricity	900×10^6 Btu or 30,000 Btu/sq.ft.

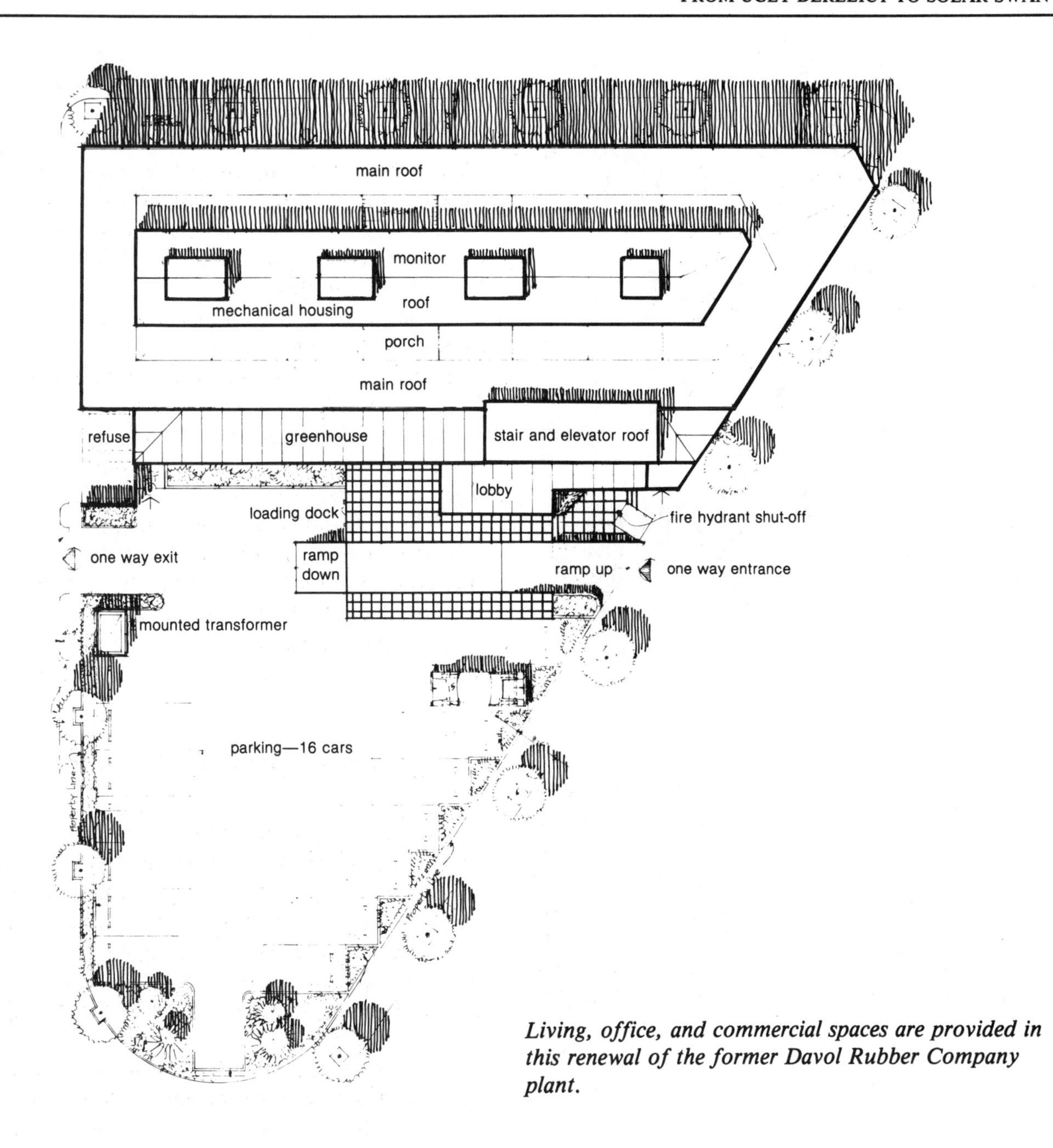

Living, office, and commercial spaces are provided in this renewal of the former Davol Rubber Company plant.

From Ugly Derelict to Solar Swan

As with the retrofit of many historic buildings, the sparkling solidity of renewed splendor does not always reveal the myriad of delays and frustration of the renewal and construction process. The 1880 Simmons Building represents the first phase of the rehabilitation of a landmark 250,000-square-foot industrial complex in downtown Providence, Rhode Island. "Davol Square," formerly the Davol Rubber Company Plant, promotes mixed use zoning in the downtown area. The complex as a whole provides a combination of living, office, and commercial spaces. The Simmons Building initiated the mixed-use concept within a single building for Providence.

East elevation. Separate entries are provided for residential and commercial spaces.

The Simmons Building, a registered landmark, incorporates planning, passive solar, and architectural concepts intended as a prototype for future historic renovation and adaptive reuse. The building has commercial rental space on the ground floor and 20 condominium residential units on the four upper floors. There are separate entryways to the residential and commercial spaces. With the addition of an interior mezzanine floor and a roof monitor floor, all but two apartment units were provided southern exposure without altering the existing street facades.

Prototype for Retrofitting

The 3000-square-foot, four-story passive solar greenhouse provides a solar collector that uses the existing building's masonry mass as thermal storage, and windows as solar apertures. Full height translucent water tubes, as unit dividers, provide additional heat storage. The greenhouse facade follows the form of the existing facade. Clear double-glazed windows are recessed into the plane between wood members that express the stone sill and lintel line of the existing wall and maintain the original 12-over-12 sash arrangement facing outside. Translucent polycarbonite plastic panels form the "solid" plane behind. Individually controlled, retractable awnings provide colorful shading control for summer months.

The new monitor floor is representative of an element found in nineteenth century mill buildings. Even so, this innovative addition cannot be seen from neighboring streets. On the Simmons Building, it serves to ventilate the fourth-floor units and provide a southern exposure for the north-facing three-story apartments.

The greenhouse and monitor together greatly enhance the quality of living spaces, reduce the energy consumption for the residents, and, most important, illustrate how solar energy concepts can be incorporated in the restoration of a historic building.

Lively Accommodations

While the ground-floor rental spaces are an open clear space awaiting tenant subdivision and custom finishing, the upper residential floors are developed into a series of lively multi-story apartments. Each apartment is unique in layout and character. On the second and third floors are 11 apartments, all entered on the second floor. All but three studio apartments have a duplex layout with rooms on two floors and internal stairs. This townhouse or maisonette arrangement allows maximum access to the solar greenhouse while also giving apartments windows along the north wall for cross-ventilation and views of the downtown financial district. The nine loft-type apartments that start on the fourth floor are even more ingenious in their three-dimensional planning. The architects have called them "triplexes" because they have accommodations on three floors with private interior stairs. A new mezzanine floor provides a sleeping balcony. With the addition of a new "monitor" floor in the center of the roof, all nine apartments also have a top room with windows facing both

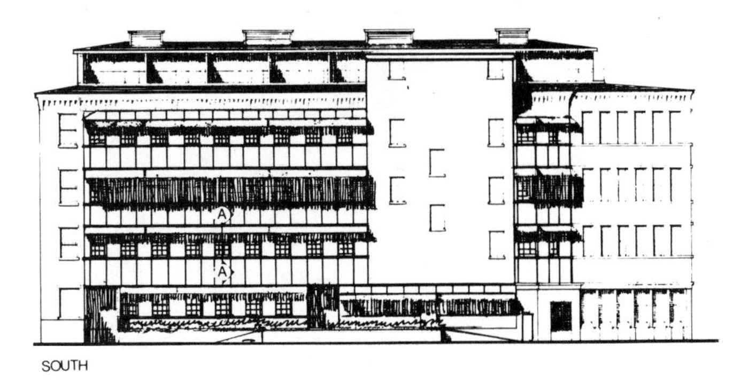

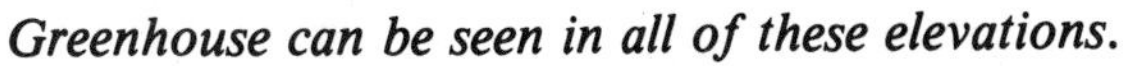

Greenhouse can be seen in all of these elevations.

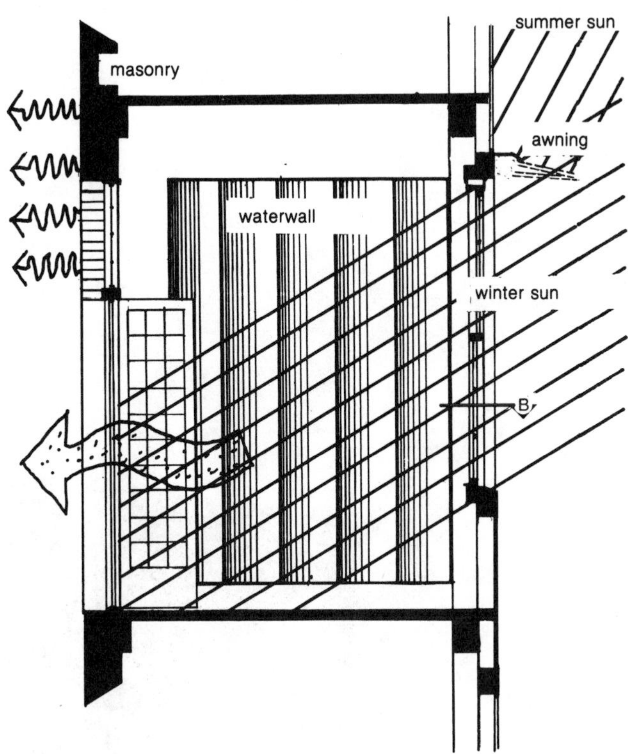

In this greenhouse elevation, awning blocks rays of sun in summer.

north and south, together with outside porches for entertainment and excellent views in all directions.

Adaptive Use

The adaptive use of this historic structure adds variety and depth in the new life of its renovation. The designer's sensitivity to the historical and visual aspects of the project brought support from unusual directions. For instance, in 1981 after the Davol Square had been in planning over two years a large architectural/engineering firm, C.E. Maguire, purchased one of the other structures in the main complex. Their complete renovation of that three-story brick building allowed occupancy early in 1982 and hastened both the financing and completion of the complex. On the other hand, the mixed use of the Simmons Building prompted a series of legal maneuvers from a neighboring property owner, the Narragansett Electric Company, which objected to the residential rezoning of the Simmons Building. The Providence Zoning Board and Rhode Island State Court upheld the "mixed use with residential" zoning three times during the process.

Although passive solar design does not totally dominate this historic renovation, it was a very supportive aspect from many quarters. Energy economics were a decisive part of the long-term financing of the project, because a healthy cash flow was assured which would not be sensitive to uncertain future energy costs. The creative architectural qualities of Davol Square inspired admiration from both protectionists and historians who publicized the wedding of new technology and old buildings. But not least is that the architects as part owners of the project have had an ideal opportunity to demonstrate their dedication to solar building and adaptive reuse in a handsome and lively remodeling that would be a prize in any city.

Interactive Girl Scout Center For Philadelphia

BUILDING

Girl Scout Program Center
Shelly Ridge
Philadelphia, Pennsylvania

DESIGNERS

Bohlin Powell Larkin Cywinski
architects/planners/engineers
Wilkes-Barre/Pittsburgh/Philadelphia
Peter Bohlin, principal
Dick Powell, principal
Frank Grauman, project architect
182 North Franklin Street
Wilkes-Barre, Pennsylvania 18701

OWNER

Greater Philadelphia Girl Scout Council
1411 Walnut Street
Philadelphia, Pennsylvania 19102

CONTRIBUTORS

Burt Hill Kosar Rittlemann Associates
solar consultants
400 Morgan Center
Butler, Pennsylvania 16001

BUILDER

Don Erb, builder
Lutz Road, RD 4
Boyertown, Pennsylvania 19512

STATUS

Occupied Summer 1983

Estimated Building Energy Performance

BUILDING DATA

Heating degree days	4,865 DD/yr.
Cooling degree days	1,104 CDD/yr.
Floor area of conditioned space	6,078 sq. ft.
Volume of conditioned space	113,185 cubic ft.
Building cost	$550,000
Cost of special energy features	$56,000
Heating	
Area of solar collection glazing	1,829 sq. ft.
Auxiliary, electric resistance	27,460 Btu/sq. ft.
Hot Water	
Need 55 gallons at 120° F. (incoming 50° F.) Solar collector to provide 70%	100 sq. ft.

ANNUAL ENERGY SUMMARY

Nature's Contribution	
Heating	59.1×10^6 Btu
Hot water	18.2×10^6 Btu
Auxiliary Purchased	
Electricity	216.3×10^6 Btu or 35,580 Btu/sq. ft.

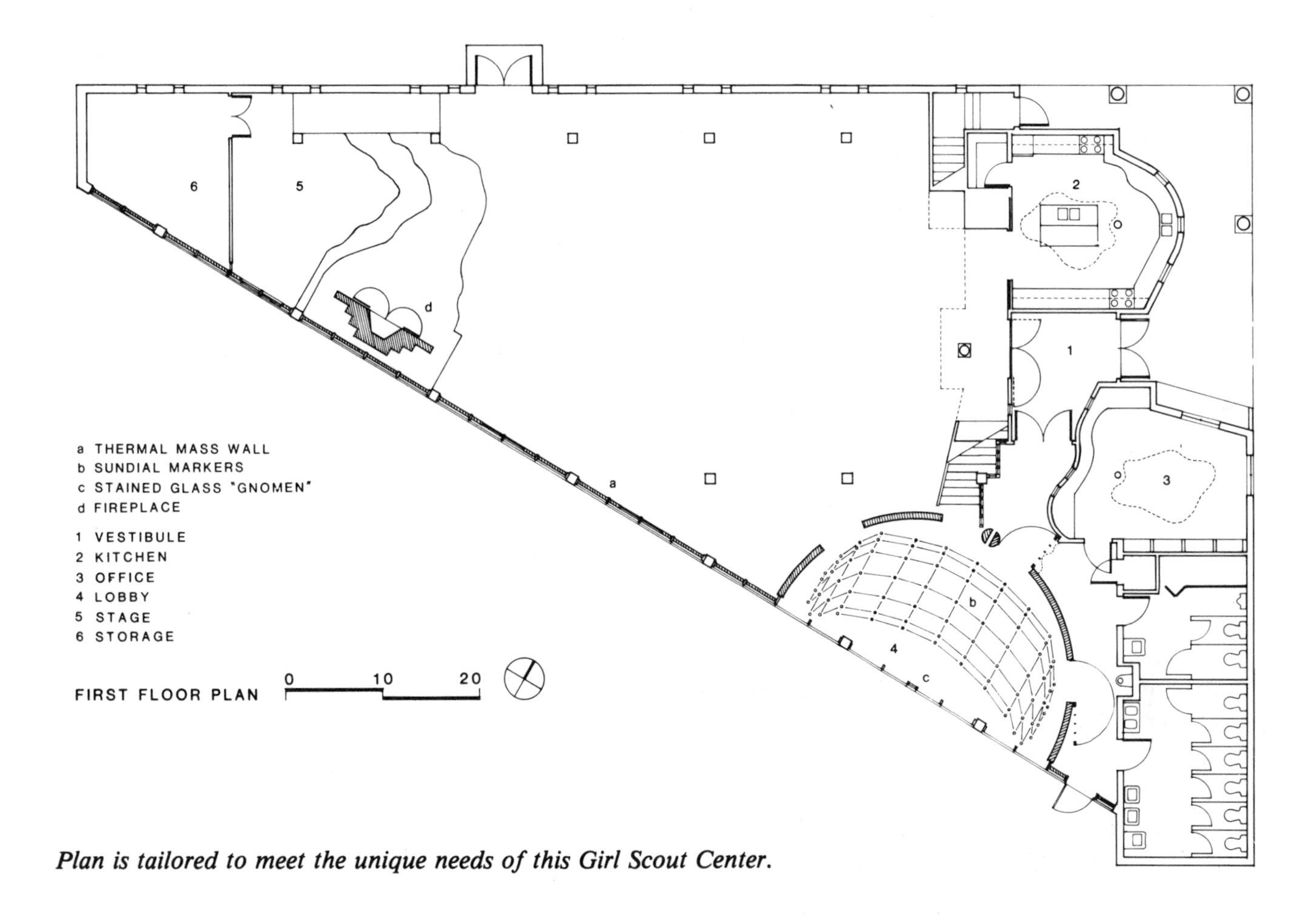

Plan is tailored to meet the unique needs of this Girl Scout Center.

The Program Center Building is the centerpiece of the Shelly Ridge Girl Scout Center, the primary facility for year-round indoor and outdoor programs for greater Philadelphia. Designed as the final construction phase in a dedicated nature preserve, this major structure satisfies the client's mandate for a building solution that would excel both in architectural quality and in energy conservation. In particular, its high visibility implied that it should be an educational device itself, as well as symbolize the Girl Scout commitment to nature and environmental issues. From the beginning, a passive design approach appeared preferable because of the possible reduction of fuel use. But even more important was the learning opportunity of a building whose systems were interactive with its users: occupant involvement in the daily operation of the facility was a primary goal.

Shaping the Architecture

To coordinate the design of the Program Center with the existing structures, a traditional architectural form in the region was continued: a pitched-roof, gable-ended building faces its junior-sized counterpart, the caretaker's house, on axis across a grassy meadow. That entrance facade establishes a formal dialogue between the building and its surroundings as well as between the various elevations of the Program Center itself.

Each side of the building is quite different. Some of that difference is generated by a triangular plan with its two points cut off. Thus the rectangular plan geometry of a conventional gable-roofed

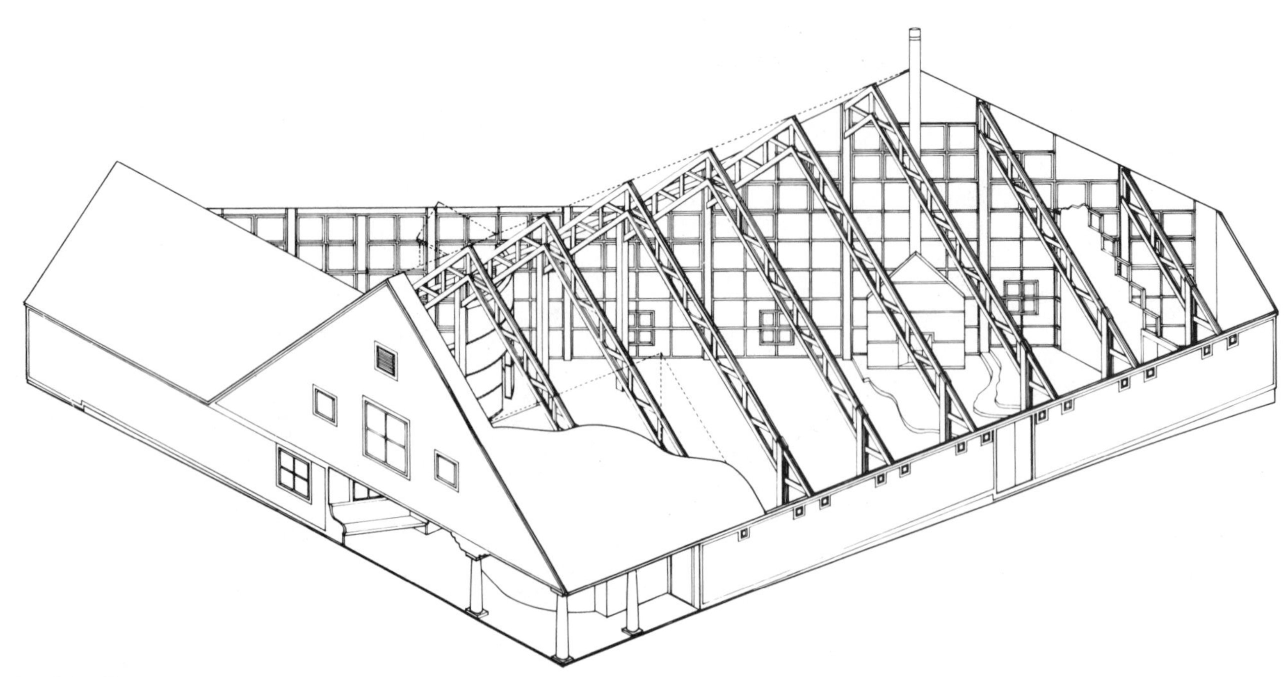

In this diagram the rear wall faces the south; thus the northeast (left) and northwest walls form an angle that points almost directly north.

building has been "sliced" on a diagonal, creating a southerly "sun face." The resulting truncated shape maximizes the available vertical solar window while minimizing the stormy northeasterly and northwesterly exposures. A line of trees and the low roof eaves on the northwesterly side protect against the winter winds, which predominate from that direction.

Celebrating the Sun

As an overall design concept, the Program Center building emphasizes and rejoices in its passive solar features, while integrating them into imaginative variations of traditional forms. Using the device of a sundial-lobby, the building's users can participate in the solar design despite the fact that the building is approached and entered from the north corner. A recessed porch leads through a double-doored vestibule to the playful solar lobby, or directly into the large activity space.

Being a 6,000-square-foot building, dominated by a single 3,000-square-foot main space, the Program Center Building is thermally influenced largely by the external energy forces of climate. This indicated that solar strategies would be most helpful in offsetting the winter heating load. In summer, Philadelphia has high humidity and a lack of diurnal temperature swing. Thus the primary opportunity for passive cooling is by ventilation. Other passive cooling approaches such as earth contact, radiation loss, or evaporation would be ineffective, but the general openness of the interior allows cross-ventilation from any direction. Because summertime activities are largely out-of-doors, artificial cooling was not required. Solar energy also heats service hot water and provides almost all the building's lighting requirements.

Because of the building's compact size, it was determined that it could be served in winter from the side by a single vertical solar collection area or sun wall. A Trombe wall approach for heating was selected for the main hall because it could function as heat collector continuously without losing its ability to be darkened for audio-visual presentations. A Trombe wall also allowed thermal storage to be easily a part of interior surfaces. The direct-gain lobby space was designed to give access to the southerly views, and to make the solar features a significant part of the entrance experience. This lobby, semicircular in plan, is designed to function as a three-dimensional sundial and to underline the building's ties to nature, while illustrating the sun's daily motion.

The direct gain sundial space is triple-glazed. It has a concrete floor and semicircular brick rear wall. A study model was used in a heliodon to confirm that this rear wall receives direct sunlight during winter months. Night insulation, in the form of a side-pulling R-5 curtain that would be manually activated, was not cost-effective when compared with triple-glazing.

The Unique Thin Trombe Wall

The main hall's Trombe wall is an unvented design. Computer runs using a TRNSYS program indicated that a vented design would produce only marginally more heat at considerable added expense. The wall consists of a structural timber grid, with 4-inch-thick brick in-fill panels. The 4-inch thickness delivers its "heat pulse" early in the day, when the building is occupied, rather than at night when it is usually empty and when the thermostat setting is lowered.

In contrast, a more conventional Trombe wall would have masonry 8–12 inches or more thick. Its heat pulse might have a time delay of 6–9 or more hours.

By using brick as an in-fill material, the unusual 4-inch Trombe thickness is structurally possible. This approach also permits the substitution of clear glazing for brick, providing a mix of direct-gain and Trombe wall areas in an integrated grid pattern in the sundial space.

The exterior glazing is a translucent plastic material in those grid quadrants of the Trombe wall that are brick-filled. This cut cost and resists vandalism. Glass is used only where a "window" to the interior exists. Manually operated insulated shades on those windows provide both day and night insulation and darkening for special events.

Because of the Trombe wall's height, fixed overhangs proved cumbersome. Moreover, the peak solar gain on south-facing walls in Philadelphia occurs, surprisingly, in September. During this month, there are only 38 heating degree days, so it is the critical time for potential overheating, but most Girl Scout activities can be scheduled outdoors. On the other hand, the solar angles of September are identical to March, when there is much more cloud cover, and when there are 716 degree days of heating required and shelter and warmth must be provided.

The design relies on conventional crank-operated, storefront awnings to overcome some of the overheating problems.

The clear areas of the sun face provide daylight to almost all spaces. High windows in the main hall face the light-colored sloping ceiling, which reflects light downwards. Daylighting is supplemented by dormers in the loft above the entrance and service spaces.

An Interesting Example of Institutional Architecture

In addition to its inclusion in the First National Passive Solar Awards program and national publication, the building was recognized by the 1981 Silver Medal from the Philadelphia Chapter of the American Institute of Architects. The AIA Silver Medal is the highest award for a project in progress.

This interesting facility defies the usual connotations of institutional architecture. Its playful forms are strongly related to the other structures and aspects of its rural setting. The boldness of its expression comes not just from a solar response but a strong architectonic sense. The low and massive northwest elevation has the timeless quality of a medieval barn. The high and open south wall with its multiple scales and single dormer window is very forward looking, with only a backward glance. Solar enthusiasts are fascinated by the life-sized sundial room which is like a walk-in cosmic clock.

New Jersey Nature Center

BUILDING

Flat Rock Brook Nature Center
Englewood, New Jersey 07632

DESIGNER

Daniel V. Scully
Equinox, Inc.
Granite Block
Peterborough, New Hampshire 03458

ASSOCIATES

Ballou-Levy-Fellgraff, architects
One Station Plaza
Ridgefield Park, New Jersey 07660

BUILDERS

Keljed Corporation
One Madison Avenue
East Rutherford, New Jersey 07073

Solair Synergy, Inc., solar subcontractor
Chatham, New Jersey 07928

CONTRIBUTORS

Winslow Fuller and Jennifer Adams,
solar system design
Total Environmental Action
Harrisville, New Hampshire 03450

Shiffman & Tietjen, mechanical engineers
Scarsdale, New York 10583
Alexanders & Von Bradskey, structural engineers
Fair Lawn, New Jersey 07410

STATUS

Occupied Spring 1980

Estimated Building Energy Performance

BUILDING DATA

Heating degree days	5,182 DD/yr.
Volume of conditioned space	43,560 cubic ft.
Net area of conditioned space	4,356 sq. ft.
Building cost	$326,000
Cost of special energy features	$36,000
Total building UA (daytime)	1,225 Btu/°F./hr.
Total building UA (night)	1,022 Btu/°F./hr.
Heating	
Net area of solar collection glazings	532 sq. ft.
Solar heat (active and passive)	50 percent/yr.
Auxiliary energy required (gas)	9,482 Btu/sq. ft./yr.

Hot Water

Need 14,600 gallons at 120° F. (incoming 52° F.)

Auxiliary—gas	1.79×10^6 Btu/yr.

Lighting

Require 1 watt/sq. ft. with 6 hrs./day

ANNUAL ENERGY SUMMARY

Nature's Contribution	
Heating	47.3×10^6
Hot water	6.5×10^6 Btu
Auxiliary Purchased	
Electricity	NA
Gas	NA

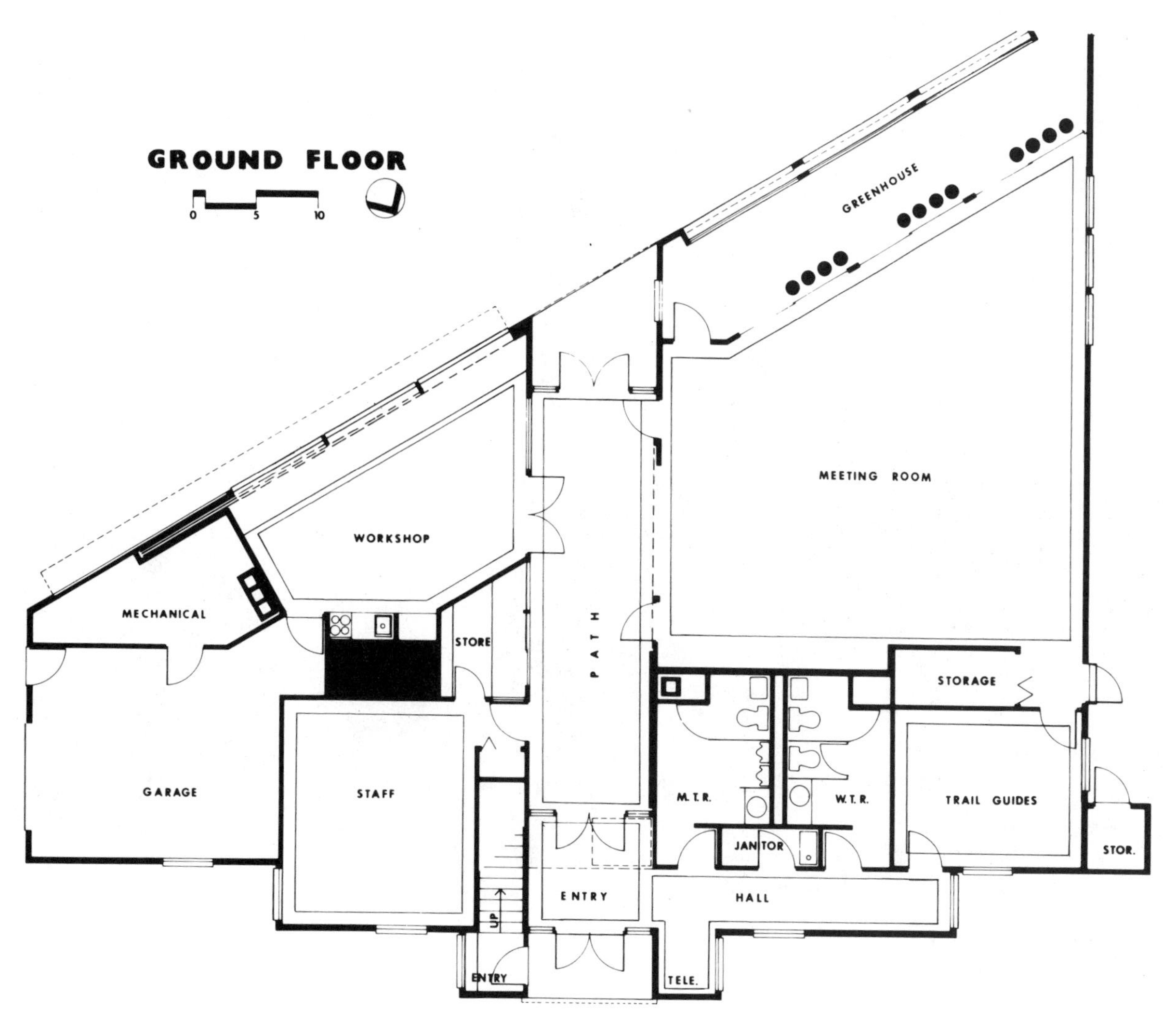

A path through the building, a greenhouse, effective uses of solar energy—all parts of this nature center.

New Jersey Nature Center

This center blends with nature in intriguing ways: it is warmed by passive solar heating, and it straddles a scenic trail in New Jersey's rocky woodland. This place in nature is a kind of node along the path, neither the beginning nor the end. Visitors belong out on the trail. If they go only to the building, they miss the center's symbolism as a place along a pathway. The 10-foot-wide trail extends straight through the building uninterrupted, forming a hall that also serves as the building's organizing element. On both sides are several simple, effective uses of solar energy. And these—in a further example of the center's educational creativity—offer lessons about passive solar energy for homeowners.

Man's Order and Nature's Order

At the Flat Rock Brook Nature Center, man's and nature's orders come together architecturally.

View from north shows entrance, at left, and, at right, one wall of the meeting room.

The north face of the building is man's formal and geometric gridded order. The south face belongs to the sun and the diagonal wall marks the solar orientation to the universe.

Man has laid a grid of roads across New Jersey. This nature facility serves in the last of the steep rocky woodland before the blocky grid of residential development spreads out across the New Jersey meadowland. Across the nature reserve land is the ghost of the same grid of once-proposed roads. These, in their present soft-edged form, are the paths throughout the center's park-like property. They survive as man's access to the natural world—to its plants, rocks, and streams.

The entrance to the center's park is like an exit ramp off the New Jersey Turnpike. Upon driving in, you face the rigid, man-made order of the north side of the building. Architecturally, the front facade builds up to the entrance, then falls back into the building. With its "reverse Mohican haircut" entrance elevation, you visually sense that, if the doors were closed, you could proceed right up and over the building. The path perseveres.

Hypotenuse of the Sun

The nature path emerges through the south facade. The path mechanically continues its arbitrary direction unaffected by the solar orientation, the mark of the sun on the building. The south face stretches out wide to expose itself—nature's hypot-

The entrance to a path through the building.

enuse of the grid-ordered plan of men. On the south face, to absorb energy, are:

- A sunspace/greenhouse with full-height windows, water-storage tubes, and insulated, garage-door shutters.
- Windows into the workshop, with a dark concrete floor slab; hollow-core door insulating shutters slide back into pockets.
- Active site-fabricated air collectors on the garage wall and on the wall above the first floor, as the architectural frieze.
- Windows across the full width of the upstairs apartment from wainscoting height to the ceiling.
- Solar domestic hot water collector on the roof.

An Elementary Plan

To the left, immediately after passing through the double doors of the entry vestibule, are the staff offices and sales shop. The toilet and coat rooms and access to the trail guides' room are on the right. Instructional spaces that are used throughout the day are located on the south side. To either side of the central hall are the public meeting and workshop spaces, which can be part of the hall or closed off and accessible only through a doorway. Access to the second-floor apartment for resident managers is through an outside entry.

Entrance on south side of building.

Substituting Sun for Oil

Without any contribution from solar energy, the Flat Rock Brook Nature Center would require approximately 2,235 gallons of oil per heating season. By using solar gains in the design of this building, the heating requirement is reduced by about 50 percent. This is achieved, not by one solar system, but by the cumulative gains of several different systems. For instance, direct gain through windows in the workshop accounts for an 11 percent contribution to the total heating, while the direct gain to the second-floor apartment contributes another 6 percent. The greenhouse wall of the main meeting room contributes 17 percent to the total. The two active wall collectors contribute another 15 percent. Together, these solar features contribute the equivalent of approximately 1,112 gallons of oil per heating season.

The workshop space has windows from counter height to ceiling and a concrete slab floor for thermal mass. Insulating shutters made of hollow-core doors slide out of their pocket in the mechanical room to cover the windows at night. The hollow-core door is excellent to use for insulating shutters. It is as effective as any shutter can really be: cheap, stock, familiar, and it has no fire code problems.

The sunspace/greenhouse, with floor-to-ceiling glass, requires more thermal mass than the concrete slab. This is achieved with 12-inch diameter water tubes. Insulated garage doors, which are effective and easy to operate, are used as thermal shutters. For protection from vandalism, the doors are located on the exterior of the glass. Temperatures inside the sunspace/greenhouse usually remain in the 55° to 75° range, so the space serves as thermal buffer between the meeting room and the cold outside air.

Storing Heat for the Night

Maximum advantage is taken of all the solar energy falling on the south wall. As the garage needs no direct heating, the south wall of the garage is converted to a simple site-fabricated solar collector. This as well as the 4-foot-high active site-fabricated solar collector above the first-floor windows, is connected with duct work to a rock bed heat-storage container between the workshop and offices. Heat collected during the day can be blown directly to the staff offices, which do not have their own south-facing windows, or it can be put into thermal storage. At night or on cloudy days, this heat can be drawn out of storage and supplied to the staff offices and toilet facilities.

The backup heating system is gas-fired hot water. The collectors are all site-built of corrugated aluminum, with air flow behind the absorber. On the collector adjacent to the windows, the glazing is glass. For the overhead collectors, long horizontal lines were desired, as well as the flexibility of corrugated sheets of plastic glazing. The inner layer of glazing is Kalwall Sunlite and the outer is corrugated Filon. The duct work between the collector,

View from the inside and looking south. Meeting room is at right.

rock bed, the heated space, and the solar air handler is exposed for visitors to marvel at. The duct work's spacey design adds character to the ceiling especially in the hallway.

Nature Education

The educational opportunities of any nature center are numerous and perhaps obvious. Here, familiar building materials and spaces communicate thermal sense on a domestic scale. Passive solar concepts incorporated in the building are appropriate to most one-family homes. But the larger size of the building and its presence as a community facility mark it with institutional authority beyond its symbolic resolution of the geometrics of man and nature. Thereby, its imagery transposes passive solar techniques from residential applications to the larger dimensions of social responsibility. In the process, it entertains while it informs.

Visitor Center For The California Poppy Reserve

BUILDING

Jane S. Pinheiro Visitor/Interpretive Center
Antelope Valley California Poppy Reserve
Antelope Buttes: Western Mojave Desert
14 miles west of Lancaster on Avenue I, California

DESIGNERS

Robert D. Colyer and S. Pearl Freeman, architects
The Colyer/Freeman Group
1945 Eddy Street
San Francisco, California 94115

OWNER

Department of Parks and Recreation
State of California

CONTRIBUTORS

Shapiro, Okino, Hom & Assoc., structural engineers
1736 Stockton Street
San Francisco, California 94133

Marion, Cerbatos and Tomasi, electrical and plumbing
2015 Steiner Street
San Francisco, California 94115

CONTRACTOR

Dermody, Inc.
Lancaster, California 93534

STATUS

Constructed 1981; opened April 1982

Estimated Building Energy Performance

BUILDING DATA

Heating degree days	2,484 DD/yr.
Cooling degree days	2,755 CDD/yr.
Net floor area of conditioned space	1,635 sq. ft.
Volume of conditioned space	15,333 cubic ft.
Building cost	$255,000
Cost of wind electric system	$22,000
Cost of other special energy features	$6,000
Total building UA daytime	490 Btu/°F./hr.
Total building UA night	346 Btu/°F./hr.
Heating	
Net area of solar-heat collection glazings	378 sq. ft.
Solar heat	100 percent
Auxiliary heat	None
Cooling	
Earth sheltering, cool tube, evaporative fountain, natural ventilation	100 percent
Auxiliary cooling	None
Hot Water	
Passive breadbox solar heater	100 percent
Lighting	
Electric energy used	1.82 Kwh/sq. ft.
Fans	
Electric energy used	.124 Kwh/sq. ft./yr.

ANNUAL ENERGY SUMMARY

Nature's Contribution	
Heating	62×10^6 Btu
Cooling	55×10^6 Btu
Hot water	3×10^6 Btu
Ventilation	Included above
Auxiliary Purchased	
Electricity	5 percent of total used = 0.8×10^6 Btu or 455 Btu/sq. ft.

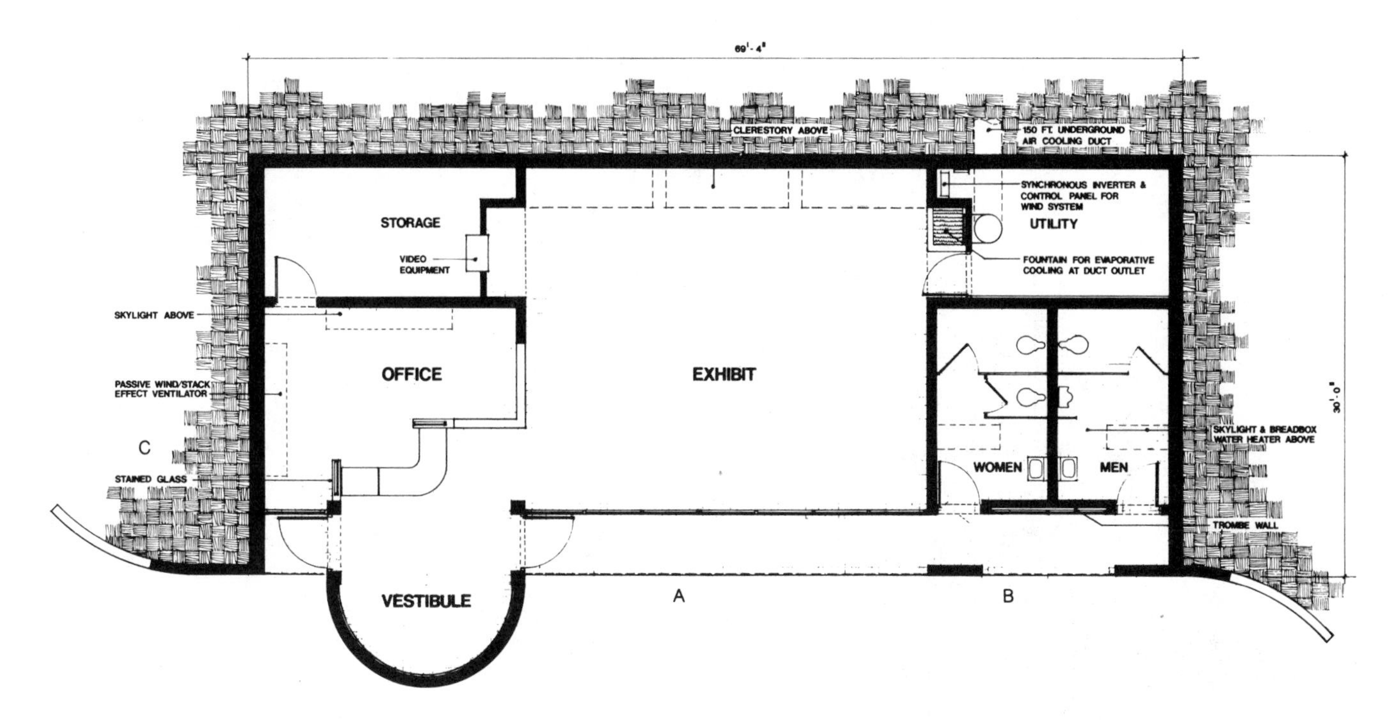

All but the south (lower) side of this building is tucked into the Mohave Desert.

From a distance only the wind generator atop a butte marks this earth-embraced visitor center in rolling fields of wild poppies and other native flowers.

The Jane S. Pinheiro Visitor/Interpretive Center is an inviting public building that deliberately takes second place to its natural environment, while exhibiting and even celebrating the wildflowers and ecology of the high Mohave Desert. The building was constructed on a 1,600-acre tract, located in the western reaches of the desert at an elevation of 3,000 feet. The acquisition of the Antelope Valley Poppy Reserve, after an eight-year campaign, and the construction of the visitor center were joint efforts of area residents, the California State Parks Foundation, and the State Parks and Recreation Department. Built for $225,000, this modern, earth-sheltered facility also serves as an example of "state-of-the-art" appropriate technology in self sufficiency. Essentially this is a zero energy building without fuel support from the distant outside world.

The Architects and Their Approach

The designers were a young architectural partnership with a combined innovative interest in energy design and community advocacy work. The California Department of Parks and Recreation became aware of their work after Robert Colyer and Pearl Freeman jointly entered and won the first prize and best-in-state award in the 1979 California Passive Solar House Design Competition. After a long search for an environmentally sensitive architectural firm with experience in passive solar design, this newly formed firm was chosen to design the visitor center. For the architects' innovative efforts, this handsome rural building of modest size has won a number of national awards for aesthetics and for energy design including the 1981 Owens-Corning Fiberglas Energy Conservation Award, in addition to its recognition in the First National Passive Solar Design Competition.

This vestibule with its stained glass windows at the sales counter dominates the south side of this building, attracting guests as they approach the center.

According to Robert Colyer, "We were, of course, thrilled and honored to receive this commission. As we developed the project, we became aware that all of the many people involved in its realization were emotionally very attached to it. The opportunity to have been involved in a project which was the dream of many, on a site of spectacular natural beauty, is both a rare and a priceless experience."

The architects' approach was initially motivated by the indigenous desert constructions of the American Southwest Indians.

Pearl Freeman added, "The historical nature and remote location of this site inspired us to design the building to be timeless. Its aesthetic expression and systems technology were to be appropriate responses that could remain independent of time, past or future."

Designed to merge into the rolling buttes while responding to the variable climatic conditions of the high desert, the earth-sheltered building has concrete-block walls and concrete roof/floor slabs. Matching the desert soil color, horizontal bands of smooth block pierce the predominantly split-face

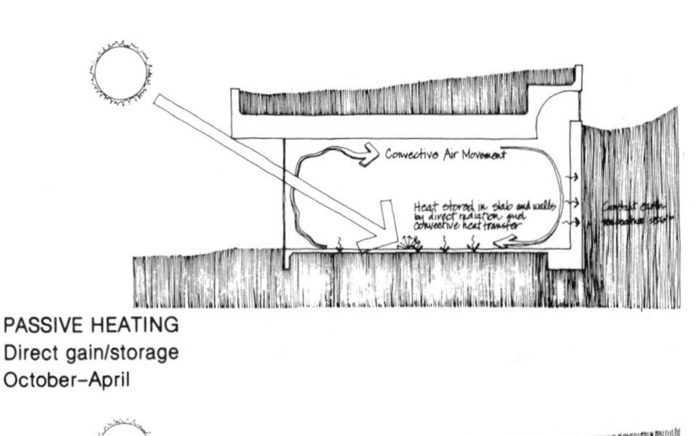

PASSIVE HEATING
Direct gain/storage
October–April

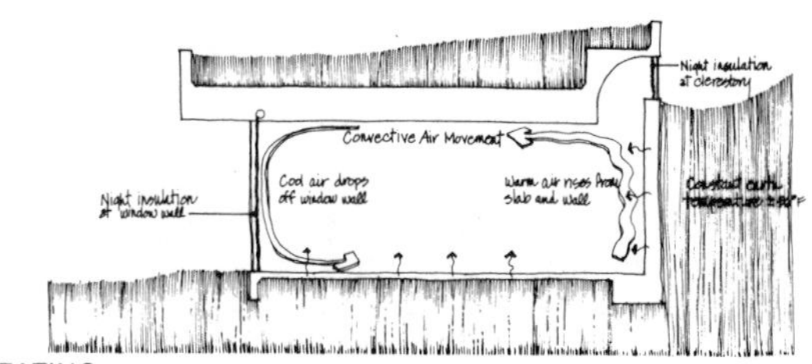

PASSIVE HEATING
Re-radiation
October–April

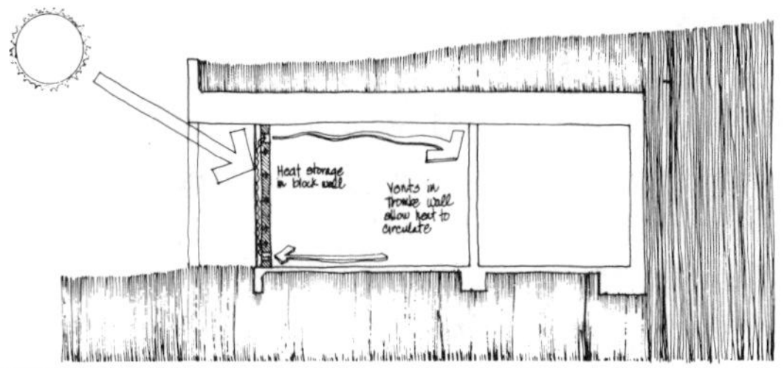

PASSIVE HEATING
Trombe wall at rest room
October–April

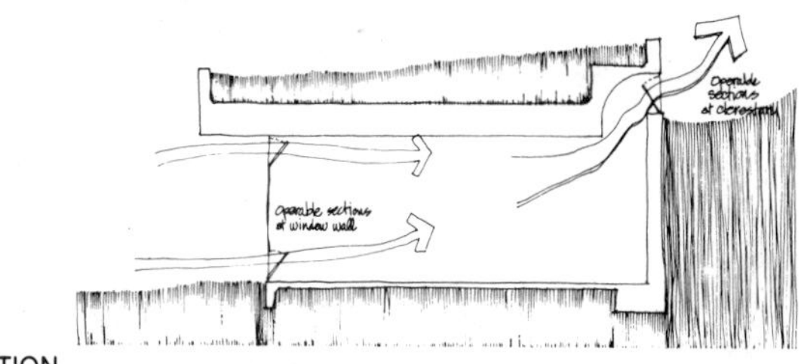

VENTILATION
Fresh air cross flow
Spring and fall

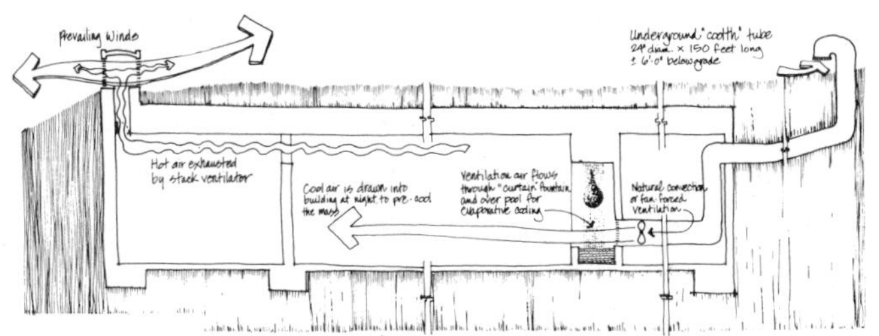

PASSIVE COOLING
Earth tube, evaporative coding fountain, passive stack ventilator
June–September

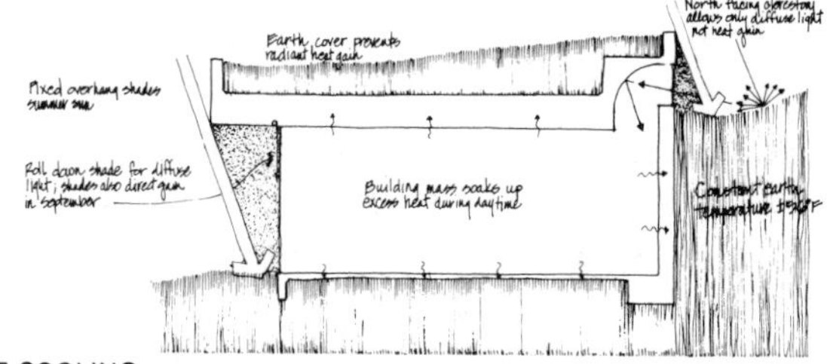

PASSIVE COOLING
Shading, mass, and natural ventilation
May–September

Passive heating, ventilation, and passive cooling—all are possible thanks to careful planning.

block wall finish, suggesting layers of native sedimentary rock. The exterior wing walls step and curve where the building meets the earth.

Responding to Nature

Both the building's form and systems respond to the remote location in a harsh climate and a fragile ecology. The underground aspect of this public building is especially crucial to energy efficiency: the earth acts as a constantly stable thermal surrounding, with average annual temperatures of about 56° F.: Earth contact moderates the extreme outdoor temperatures which reach highs of 113° F. in the summer, and lows of 3° F. in the winter. But the precise cooling contribution is difficult to predict.

Although the building is partially underground, the interior is daylit and airy. Intended primarily for day use only, the visitor center is equipped with fixtures for supplemental and night lighting. Strip fluorescent lighting in the skylights and clerestory simulate daylight, and incandescent track lighting and recessed incandescent fixtures handle nighttime needs. Annually, only 1.66 watts per square foot of artificial lighting is required. With a 30-foot north clerestory window along the rear wall, daylighting is the primary source of light in the exhibit room. It does not introduce signifi-

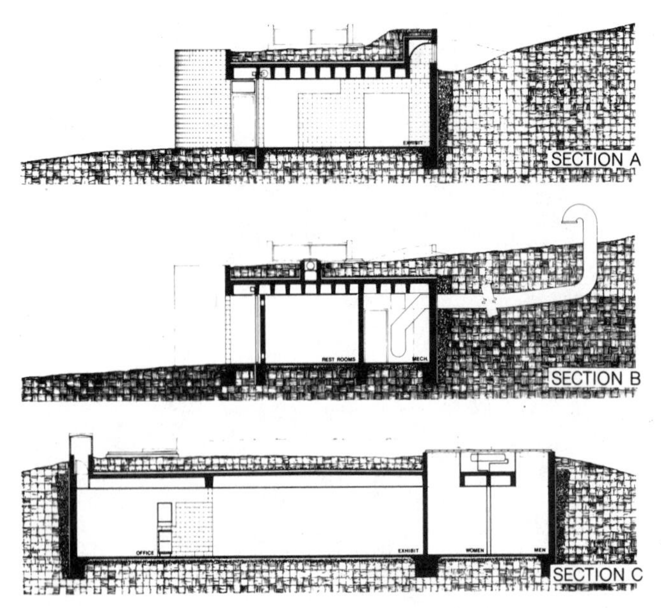

Sections A, B, and C, as indicated on floor plan, page 217.

cant heat gain. The symmetrical niches flanking the daylighting rear wall frame two of the center's permanent displays, the video exhibit and the evaporative cooling fountain.

Daylighting is also the major source of lighting in the office/vestibule area. Backed by a skylight and floor-to-ceiling windows, the stained glass windows at the sales counter sparkle dramatically as one enters the building. The stepped/eroded wall in the same area visually unites the two spaces while opening the spaces for the passage of ventilation air and daylight from the exhibit room. The 2,000-square-foot building can be expanded from either the east or west without violating the original design concepts.

A Natural Energy Demonstration

This completely self-contained demonstration project demanded natural energy systems that were both passive and observable like its desert predecessors. The visitor center uses the sun's energy, the vigorous winds, and the stable temperatures of the earth to provide its heating, hot water, electricity, and cooling needs. Using no back-up heating source, 100 percent of the building's heat comes from direct gain-mass storage in the exhibit/office vestibule areas and a vented 8-inch concrete block Trombe wall in the restrooms.

Passive solar water heating was considered appropriate for the building's relatively minor hot water requirements. A breadbox unit, basically a black tank, was placed under the skylight between the two restrooms.

Computer analysis using a CALPAS program showed that during the winter months, using R-5 night insulation on the glazed areas, the building will maintain about 67° F. during visitor hours. After the first winter of the building's operation, the park ranger, who informally monitored the performance, expressed satisfaction with the center's thermal comfort levels. He noted that the interior temperature fell to 62° F. only once that season, during an unusual five-day overcast period that included a snowfall of eighteen inches.

The exhibit areas are cooled in the summer by nighttime pre-cooling of the mass using a 150-foot-long underground duct, assisted by a 1/3 horsepower, 2,200 CFM fan.

Cooler night air is first funneled through the 7-foot-deep duct, releasing several degrees of heat to the surrounding earth. As it enters the exhibit area, the air passes over an evaporative cooling fountain to increase its humidity and thereby cool it further.

At the opposite side of the building, hot air is passively exhausted by a convection stack ventilator. With the low relative humidity range of 11 to 24 percent augmented by evaporative cooling, preliminary analysis showed that a 15° - 20° temperature differential to the outside air can be expected. Cooling is further enhanced by an overhang and roll-down shades of 12 percent transmittance. Experience to date has confirmed these estimates. Electricity for lights and equipment such as motor operators for night insulation and stack ventilator louvres is provided by an 8K Jacobs wind-electric generating system. An electrical tie with the public utility, Southern California Edison, uses a synchronous inverter, so that surplus electricity can be used by the utility and later borrowed back. Lo-

This 8K Jacobs wind-electric generating system marks site of center.

cated in one of the windiest parts of the state, the wind generator system is expected to pay for itself in 11 years.

The building is structured to display the internal workings of its natural energy systems. Items like the evaporative cooling fountain, south glass and masonry heat storage, natural ventilation systems, and wind-electric generator are part of the experience of visiting this park. After viewing the exhibits and the building systems, the visitor continues on the hiking trail to walk among the wildflowers, while, in the distance, the building becomes an inconspicuous part of the butte, and a sophisticated and intelligent extension of nature.

The visitor center was constructed during the summer of 1981, with an opening date of April 18, 1982, for the wildflower season. This demonstration project is an innovative example of an energy-efficient building that is especially important because it is visited by tens of thousands of wildflower admirers each spring. The design reflects the architects' primary goal of integrating aesthetic formalism with "state-of-the-art" appropriate technology. The client, an agency of state government which has long been in the forefront of the development of energy-efficient buildings, deserves recognition for the creative planning vision of this pleasant and intimate institutional building isolated in the desert. Most appropriately it honors the state flower, the California wild poppy.

Library Addition In Historic Massachusetts

BUILDING

Richard Salter Storrs Library Addition
Historic District
Longmeadow, Massachusetts 01106

DESIGNERS

Perry, Dean, Rogers & Partners: Architects
177 Milk Street
Boston, Massachusetts 02109

CONSULTING ENGINEERS

McCarron, Hufnagle & Vegkley Associates
340 Park Square Building
Boston, Massachusetts 02116

STATUS

Unbuilt

Estimated Building Energy Performance

BUILDING DATA

Heating degree days	6,235 DD/yr.
Net new conditioned area	11,614 sq. ft.
Volume of new conditioned space	137,000 cubic ft.
Volume of existing conditioned space	84,000 cubic ft.
Total new building UA	3,790 Btu/°F./hr.
Heating	
Net area of solar collection	2,673 sq. ft.
Auxiliary needed (No. 2 oil)	52,655 Btu/sq. ft./yr.
HVAC Fans	
Required 2.1 Kwh/sq. ft/yr.	7,188 Btu/sq. ft./yr.

ANNUAL ENERGY SUMMARY

	Nature's Contribution
Heating	425.9×10^6 Btu

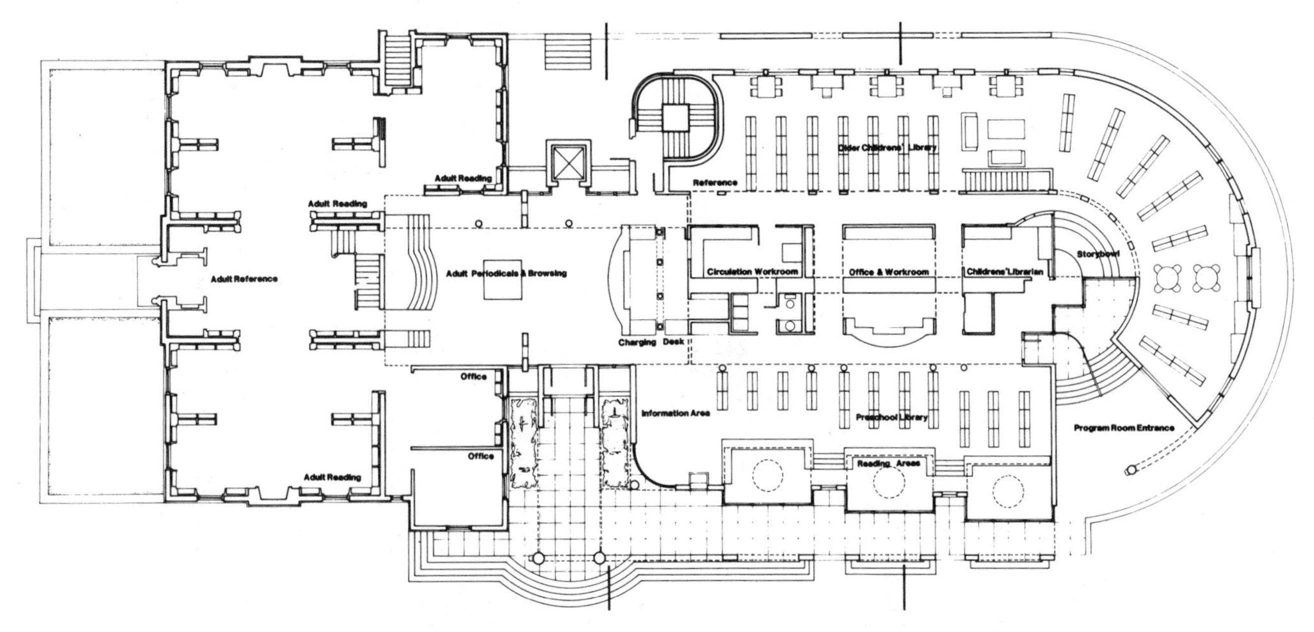

The addition to this library is all of building in front and to the right of entrance, in lower part of plan.

Designing an energy-conscious building within a historic district is especially challenging. This project for an addition to a small public library tripled the size of the original Georgian Revival building. At the same time the addition conformed to the requirements of the Longmeadow historic district, it made optimum use of solar energy and natural ventilation. By extending the addition axially to the rear of the site, it was possible both to preserve the formal front and sides of the existing building, and to provide a broad south-facing flank for the development of passive solar potential. Whereas the existing library has 6,780 square feet divided among three floors, the new wing contains an additional 12,800 square feet, primarily for children's services.

Respecting the Original Building

The original library is organized with reading rooms to the right and left of a central foyer, according to the traditional disposition of rooms in both houses and public buildings since Colonial days. The organization of the new addition respects the original plan type by extending the central foyer as a zone 140 feet to the south. This zone contains the new main charging desk and circulation space, and is expressed as an extruded gable form—the spine of the new solar addition. A new entrance is located perpendicular to this spine, at the glazed link between the original building and the addition. The new children's library is designed with components both to the north and south of the central spine. The building's cross-section was generated by the desire to articulate these library functions as well as to provide for passive solar design and daylighting of all new areas on the main floor.

Developing the Site

The scheme maintains the traditional entrance of the building and the existing site circulation with some modifications. The one-way driveway, which passes the building on the south, leads to the parking lot at the rear, and then returns around the building to the street on the north. The parking lot is expanded to accommodate the anticipated in-

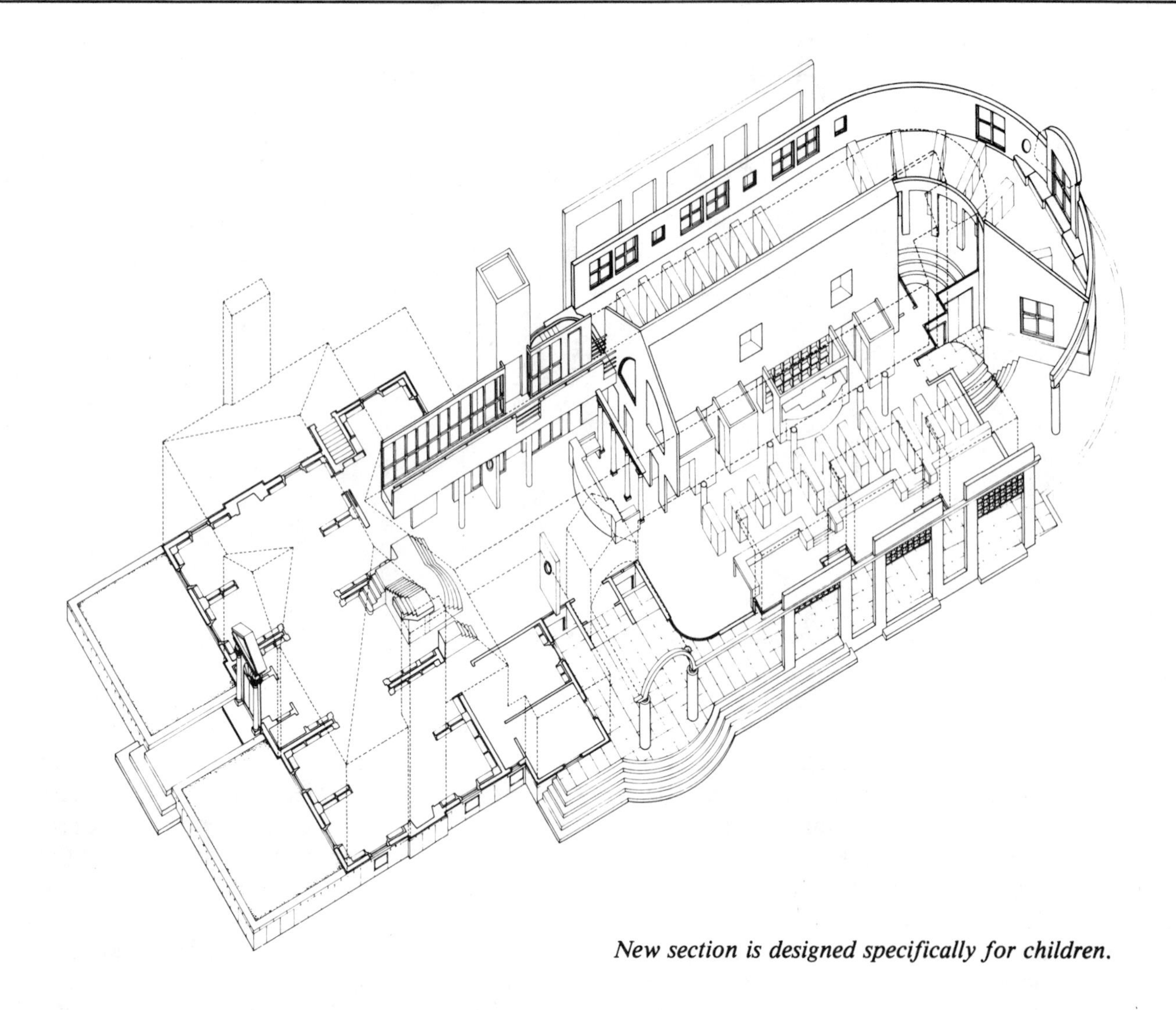

New section is designed specifically for children.

crease in the use of the facility. Leaving the existing entrance as is, the new design provides a second and more practical entrance in the form of a loggia on the south side of the addition. This covered drop-off area connects the expanded parking area with the new central browsing hall.

Respecting the Materials in a New Context

In a conscious attempt to relate the new addition to the existing library, consistent materials were proposed. White painted brick, slate paving, white wood trim, and black asphalt shingles continue the vocabulary of building materials from the old to the new. Modern energy-conserving construction, including double glass and well insulated envelope, brings a new energy standard to the look of the old. Interior painted brick walls and tiled floors provide thermal storage.

Building Form to Respect Natural Energy

The three-dimensional drawing demonstrates how the functional design of the building integrates solar energy and natural ventilation. The preschool library to the south has a flat roof form. The older

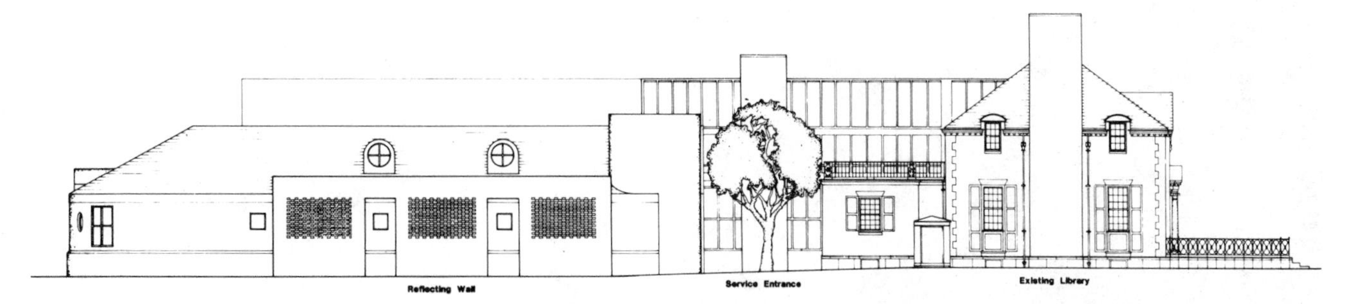

North elevation shows the addition with its center service entrance, and, at right, the original library.

children's library under the pitched roof introduces clerestory lighting from the north. The solar heat gain from the south is radiantly delivered by the mass of a freestanding thermal storage wall, a variant of the Trombe wall.

The central gabled roof form is glazed with insulating glass on the pitch facing south along the entire length of the building. A system of exterior louvres, as well as night insulation louvres on the inside, have been arranged to direct and control solar heat gain to the modified Trombe wall. The wall acts as a formal organizer along the central spine of the building as well as a gentle heat radiator.

The loggia on the south side is designed to provide 100 percent summer, noon-time shading to the fenestration of the children's space. To the north of the building the freestanding garden wall reflects south light back into the building through windows on the north wall. The brick reflecting wall, painted white on the outside to blend with the remainder of the exterior of the building and yellow on its southern concealed face, will provide warm natural reflected light for readers' spaces directly inside.

Heating Systems and Energy Management

The existing heating and conditioning systems of the old building will be rehabilitated. The old building's primary heating system will be converted from steam to hot water. In the perimeter of the new addition the basic heating system will require baseboard radiation units.

Stratified warmer air from both solar gain and conventional heating is removed from the high point of the new addition by means of exposed duct work at the top of the Trombe wall. This heat is resupplied in the basement and redistributed to supplement the convection and radiation heating systems of the building.

It is estimated that up to one-third of the heating requirements of the building will be supplied by the passive solar design. Natural lighting will replace artificial lighting throughout the year for 90 percent of the occupied spaces within the addition.

A Model That Is a Model

The architectural appropriateness of this complex solution to the demanding challenge of solar addition to a historic building is best appreciated in a handsome model that demonstrates how it would look when built. The addition, while completely respecting the past, is clearly a design that comes from the late twentieth century. It enhances both the appearance and usefulness of a century-old monument of civility. Without cuteness or self-consciousness, the integrity of the new is derived from the three-dimensional resolution of passive heating, cooling, and daylight concerns. In the architectural composition of these environmental disciplines as well as in the fresh planning and aesthetic concord, this unbuilt project is also a model of design skill in adaptive additions.

Vocational Technical Facility For University Of Minnesota

BUILDING

Vocational Technical Educational Facility
St. Paul Campus
University of Minnesota
St. Paul, Minnesota

ARCHITECT

Architectural Alliance
400 Clifton Avenue South
Minneapolis, Minnesota 55403

PROJECT TEAM

Herbert A. Ketcham, principal-in-charge; Thomas deAngelo, principal designer; Chris Johnson, Peter Pfister, Thomas deAngelo, energy concepts; Art Yellin, Linda McCracken-Hunt, George Stevens, project team

CONTRACTOR

Knutson Construction
17 Washington Avenue North
Minneapolis, Minnesota 55401

MECHANICAL AND ELECTRICAL ENGINEERS

Lundquist, Wilmar, Schultz & Martin
Baker Court, Suite 300
821 Raymond Avenue
St. Paul, Minnesota 55114

STRUCTURAL ENGINEERS

Meyer Borgman & Johnson
810 Plymouth Building
Minneapolis, Minnesota 55402

STATUS

Occupied Spring 1983

Estimated Building Energy Performance

BUILDING DATA

Heating degree days	8,196 DD/yr.
Net floor area of conditioned space	97,000 sq. ft.
Building cost ($52/sq. ft.)	$5,194,350
Cost of special energy features	$213,000
Total building UA, daytime	16,200 Btu/°F./hr.
Total building UA, night	14,000 Btu/°F./hr.

Heating

Net area of solar-heat collection glazing	4,420 sq. ft.
Auxiliary heat, below grade—low pressure steam coils and external air bypass	
Auxiliary heat above grade—hydronic	
Cost of auxiliary heat	$10,500/yr.

Cooling

Natural cross-ventilation and movable insulation.
Auxiliary, centrifugal refrigeration 250 tons.

ANNUAL ENERGY SUMMARY

Not available

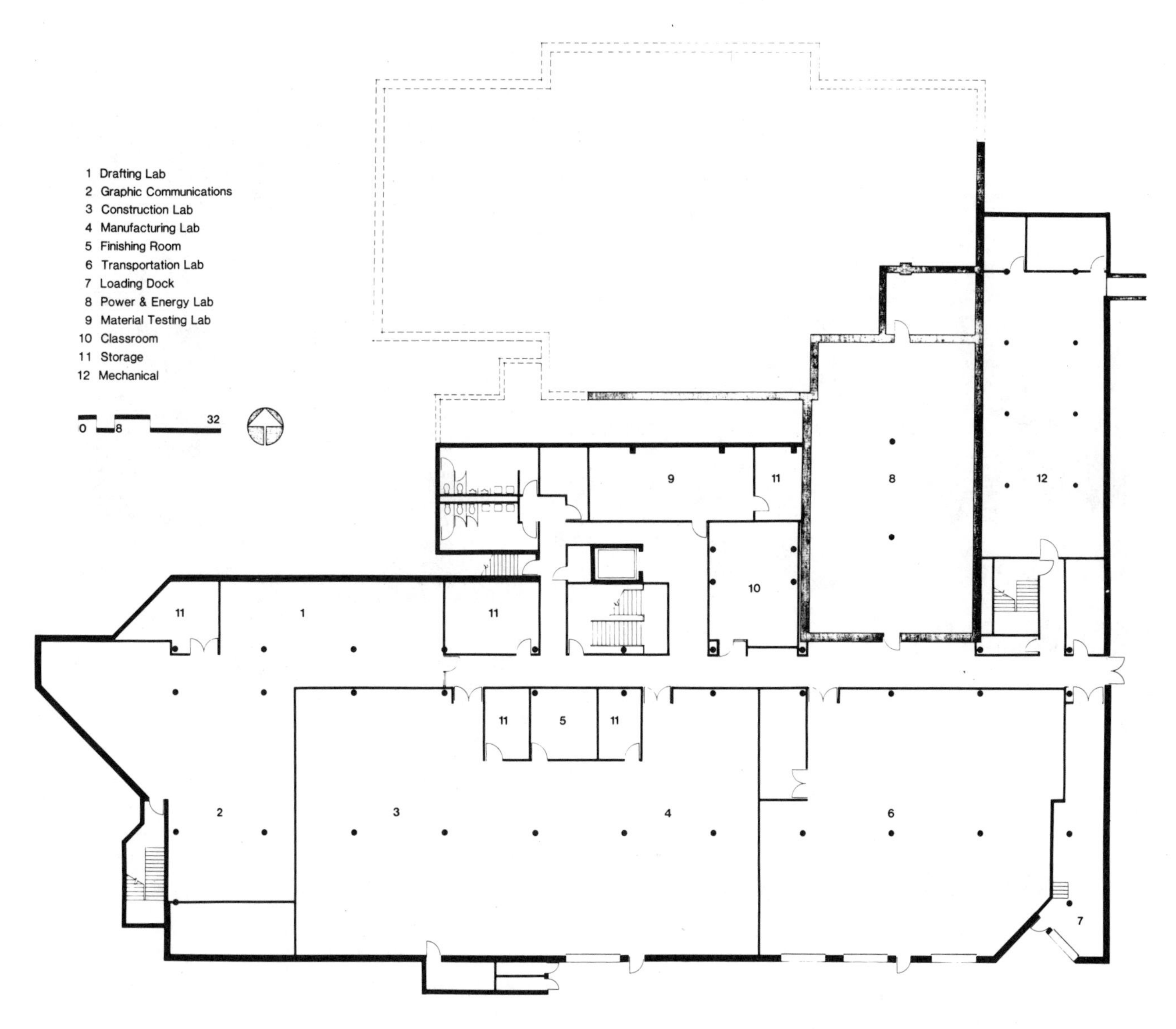

In this first-floor plan, original building outline is shown at top, with dotted line.

A new vocational and technical building is not just an opportunity for a living, educational demonstration of current building technology. It can also be a mandate for improving the energy performance as well as the functional accommodation of an existing campus. At the St. Paul Campus of the University of Minnesota, both objectives were fulfilled in the exemplary Vocational Technical Educational Facility. By linking with existing older buildings, an enclosed pedestrian concourse system was extended to provide added protection for students in the harsh Minnesota winters. Completed in the fall of 1982 at a cost of more than $5 million, the complex brings together with logic and sense an institutional community of classrooms, laboratories, and supporting facilities for six divisions of the university's Department of Vocational and Technical Education.

The original Livestock Pavilion, now expanded and converted into a vocational technical building.

Tying Together Old and New Buildings

The project included the renovation of a handsome 30,000-square-foot brick and stone Livestock Pavilion, built in 1904, and 67,000 square feet of new construction directly adjacent to and overlapping the pavilion. The design affirms the architect's belief that retaining the image of a single educational facility, composed of new and old interrelated parts, is more appropriate than creating a formal contrast between new and old. Even so, a crisp new solar wall almost 200 feet long dominates the south edge of the complex.

The visual character of the new facility is derived by uniting new and existing buildings, using both form and surface to achieve an attractive continuous composition. Two time periods of architecture have been allowed to overlap; the two brick structures are integrated by common brick and mortar colors, clear glass, and a common concrete "base" which creates a unifying podium. The base facilitates pedestrian approaches at the building entries and provides a buffer to vehicular traffic on the north and east. The south elevation is developed as a kind of passive solar greenhouse with movable insulation. In addition to its energy benefits, this facade provides a lively and transparent edge to the open space on the south, striking a contrast with the heavy appearance of campus buildings to the west and north.

The Central Atrium

These various old and new building elements come together in the central atrium of the new building. Developed as a part of the pedestrian concourse, the atrium provides a vantage point from which people can orient themselves to the organization of the building and the layout of the campus. The atrium is a student lounge, providing space for the overflow of people during the change of classes. Daylight from the atrium lights the building's office spaces. The atrium's interior design creates the mood of an exterior space. The color schemes that identify departments are derived from building material colors as revealed by a mix of natural and artificial light. Natural light enhances their interdependencies, and the subsequent derivation of interior design emphasizes that continuity. The pedestrian spine and the central space are part of a system larger than the building itself, thereby integrating "campus" identity with "building" identity. By reducing the apparent size of the building program, a balance is created with neighboring buildings.

The central atrium, sunlighted, dramatic, a place for many uses.

Integrating Passive Design Systems

The Vocational Technical Education Facility incorporates passive design strategies in two major features, the circulation atrium and the Trombe wall. Both features provide energy savings and aesthetic amenity for the building users.

The atrium and circulation space create an interior that feels and functions like an exterior space. The three-story volume is daylighted through the extensive use of glass brick and clerestory glazing. The common wall between the atrium space and the departmental offices on the south is a translucent glass block screen that allows a sharing of light and view, yet meets the required fire separation.

The atrium and circulation space are heated and cooled only by air rejected from the classroom spaces to the north and the departmental offices to the south. Fire dampered air grills between these spaces allow exhaust air from the occupied spaces to be circulated through the atrium. "The atrium circulation space is a major amenity for this campus building, yet its maintenance costs for heating, cooling, and lighting are minimal," says Tom deAngelo, lead designer on the project.

The second major passive energy feature is the Trombe wall that composes the south wall of the building. The genesis of this wall was as much aesthetic as energy-saving. The single-glazed greenhouse enclosure provides a soft, open contrast to the massiveness of the surrounding brick buildings. The solar-heated space of the Trombe wall was originally conceived as an extension of faculty offices. The intent was that faculty could have physical access to the space via operable interior windows. However, fire regulations mandated the use of fixed glazing, so this aspect was not realized. As designed and constructed, the Trombe wall provides a lively, animated, and dynamic response to its environment.

The wall is composed of an interior thermal and an exterior weather wall. The interior wall is constructed of 8-inch filled concrete block from the floor line to the window sill, a 4-foot-high single-glazed, fixed ribbon window, and a structural concrete spandrel from window head to the concrete floor slab. Both the glazing and the mass wall are

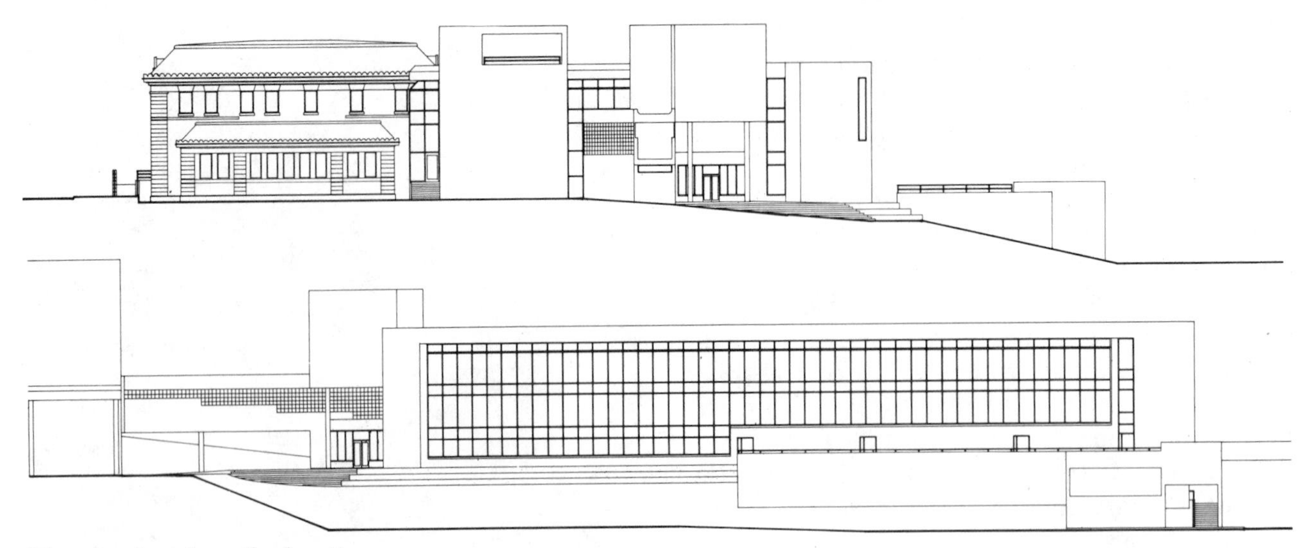

West (top) and south elevations.

insulated at night and on cloudy days by a motorized insulated patented curtain composed of four layers of reflective mylar film to create insulating dead air spaces. The perforated metal deck that is cantilevered from the concrete spandrel at the window head provides shading of the interior glass in summer and provides a catwalk for maintenance work.

The exterior weather wall is a single-glazed curtain wall system that provides moisture and wind protection for the Trombe wall components. "We realized that the traditional way to design a Trombe wall was to have no more than a 6- or 8-inch space between the mass and the glass to minimize air convection, but our concern for maintenance—primarily window cleaning and adjustments to the insulating curtains—led us to the 3-foot-deep solar heated space. Besides, the use of the insulating curtains in conjunction with the mass wall minimizes the convection problem, and gives us the best of both worlds," noted Peter Pfister, AIA, who worked with the project team to develop the energy strategies.

The Trombe wall was not vented because calculations showed that internal gains and the direct solar gains would provide sufficient daytime heat, so the design strategy was to provide nighttime heating while minimizing losses. In addition, the fire dampers required for venting would have been complex and inordinately expensive. As designed, the Trombe wall assembly cost about $200,000 more to build than a conventional insulated brick cavity wall with ribbon windows. Although space conditioning costs from the university steam plant are extremely low, it is projected that the energy savings will more than pay for the additional construction costs of the solar wall over the lifetime of the system.

Operating the Trombe Wall Assembly

The movable insulation system has a summer operation mode and a winter operation mode described as follows:

Winter day: low-angle winter sun is not blocked by sunshades, and it heats and lights office interiors; insulation is automatically raised to allow sunlight to heat masonry and concrete mass wall; on cloudy days, insulation is lowered over mass walls to keep heat in while leaving windows open to view.

Winter night: insulation is lowered over glazing and thermal mass areas to reduce heat loss.

Summer day: horizontal sunshades screen high-angled sun from interior; insulation is lowered over mass wall to minimize solar gain.

Summer night: insulation is fully retracted; greenhouse space between exterior moisture wall and interior thermal wall is vented directly to outside to provide convective cooling of interior thermal wall.

In addition, a time clock control fixes the time when the curtains respond to the solar controls. For example, on a winter day the time clock closes the curtains between 6 p.m. and 6 a.m. (or whatever time desired). During the daylight hours the curtains are controlled by solar and thermal sensors. On a cold day with little solar insulation, the curtains will stop at the window head to cover the masonry wall but will leave the glass open for daylighting and views. If the sun shines strongly and the temperature in the space increases, the curtains will fully retract to expose the Trombe wall.

Sequence of Operation for Trombe Wall Controls

An automatic controller to manage the position of the insulated curtain wall and ambient air temperature has the following characteristics:

1. A day-night auto switch and a winter-summer auto switch provide the movable insulated curtain wall control as described below:

Switch Positions	*Curtain Wall Position*
Summer/Day	Down to end switch.
Summer/Night	100 percent up.
Winter/Day	A) When the solar sensor indicates the existence of a 10 percent difference in temperature (adjustable) or more between the solar sensor and the shaded ambient sensor, the curtain should be 100 percent up. B) When the solar sensor indicates that there is less than 10 percent temperature difference between the solar sensor and the shaded ambient sensor, the curtain should be down to expose only the windows.
Winter/Night	100 percent down.

2. A time clock in the temperature control cabinet automatically indexes the controls from the day-to-night and night-to-day cycle of operation when the day-night auto switch is in the auto position.

3. A remote bulb thermostat set at 50° F. outdoor temperature automatically indexes the controls from summer to winter and winter to summer when the winter-summer auto switch is in the auto position.

4. Adjustable time delay relays maintain the curtains in position until the solar sensing equipment has the time required to detect the solar effects.

Controls for the ambient air temperature have the following characteristics:

- When the summer-winter auto switch is set for summer operation, a room thermostat modulates the nine fresh air dampers and cycle PRV's #1 thru #9 to maintain space temperatures. When the switch calls for winter operation, the fresh air dampers close and the PRV's are off.
- The fresh air dampers open and the PRV's start if the smoke detector senses products of combustion in the space regardless of the summer-winter auto switch position.

Thermal Logic and Architectural Expression

The new building is not a dazzling or eccentric expression of a new solar world, rather it is a sensible and handsome background to educational functions with design logic and technical good sense.

Extending and elaborating the enclosed pedestrial network of the campus was a logical spatial backbone in organizing a complexity of interior needs. But the long solar wall is the most visible expression of exterior architectural continuity.

Both of these major architectural features have critical roles in the passive delivery of comfort. The daylit atrium is a social forum and a daylight source that also operates as an air plenum.

Expanding Resources For A Small College

PROJECT

Site Two Development
Atlantic Community College
Mays Landing, New Jersey 08330

DESIGNER

CUH2A
45 State Road
Princeton, New Jersey 08540
Harrison J. Uhl, Jr., partner in charge
Charles Kronk, project architect
Joseph Scopino, designer
S. Louis Kelter, mechanical engineer
Ronald F. O'Brien, electrical engineer

CONSULTANT

Dan Nall
Berkeley Solar Group
3026 Shattuck Ave.
Berkeley, California 94705

BUILDER

Arthur J. Ogren, Inc.
210 E. Garden Road
Vineland, New Jersey 08360

STATUS

Completed 1982

Estimated Building Energy Performance

BUILDING DATA

Heating degree days	4,946 DD/yr.
Cooling degree days	864 CDD/yr.
Net floor area of conditioned space	14,600 sq. ft.
Volume of conditioned space	157,700 cubic ft.
Building cost	$1,472,000
Cost of special energy features	$63,000
Total building UA, daytime	2,169 Btu/°F./hr.
Total building UA, night	1,800 Btu/°F./hr.
Heating	
Net area of solar-heat collection glazings	1,954 sq. ft.
Auxiliary heat needed, natural gas	10,320 Btu/sq. ft./yr.
Cost of auxiliary heat	4.1¢/sq.ft./yr.
Cooling	
Cross ventilation and natural convection	
Auxiliary cooling needed, electric refrig.	2,106 Btu/sq. ft./yr.
Cost of auxiliary cooling	5.2¢/sq. ft./yr.

Hot water

Need 17,280 gallons at 110° F. (incoming 50°F.)

Cost of fuel, electricity	$112/yr.
Lighting	
Electric energy used	3.188 Kwh/sq. ft./yr.
Cost of auxiliary lighting	27¢/sq. ft./yr.

ANNUAL ENERGY SUMMARY

Nature's Contribution	
Heating	37.1×10^6 Btu
Cooling	43.5×10^6 Btu
Daylighting	63.8×10^6 Btu
Auxiliary Purchased	
Electricity	Not available
Natural Gas	Not available

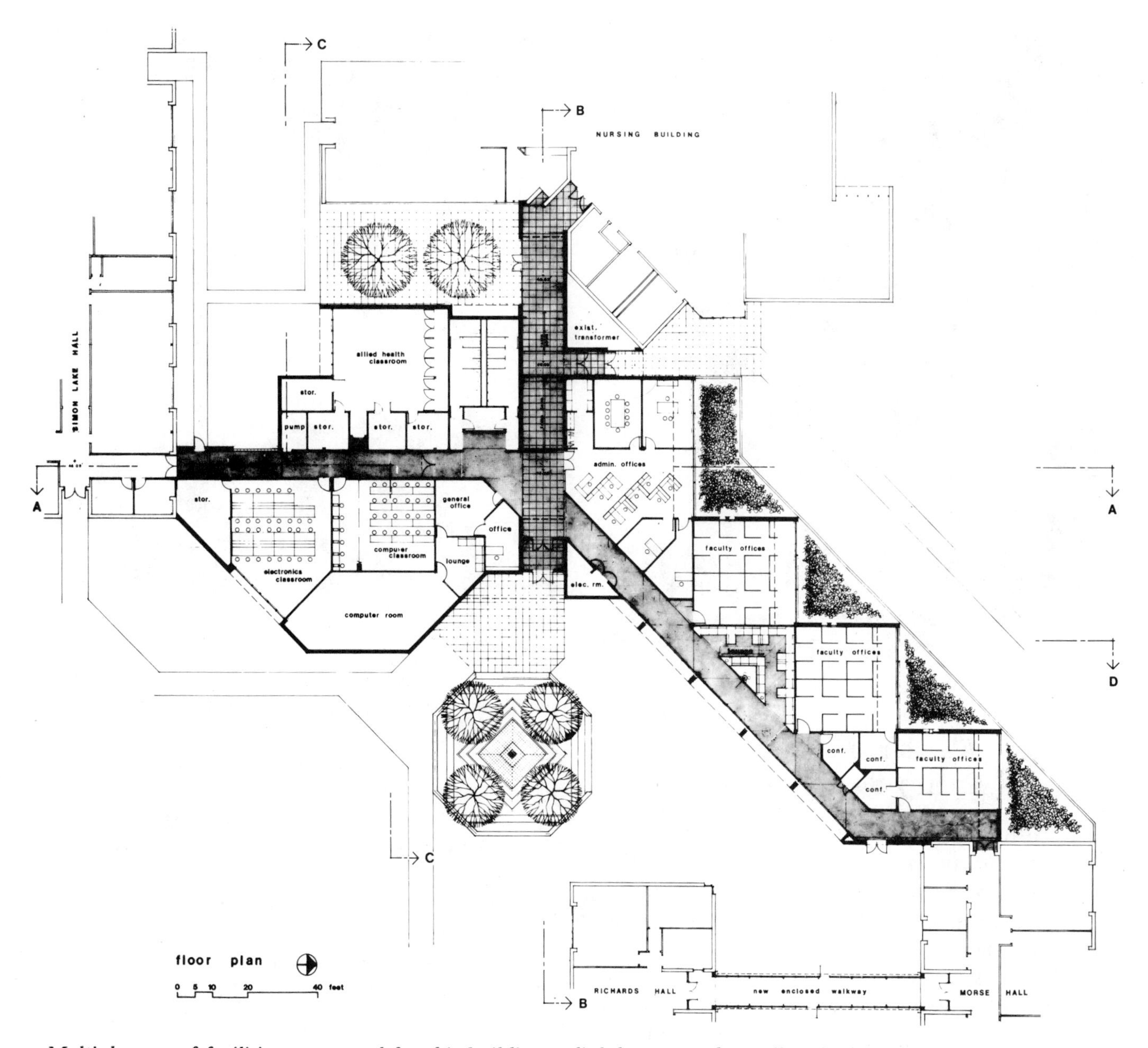

Multiple uses of facilities was a goal for this building, a link between other college buildings.

In 1979, there were more than 4,000 students at Atlantic Community College, a liberal arts school in New Jersey with a campus designed for 800. This overcrowded condition forced college authorities to house many programs temporarily in trailers. Moreover, it required an immediate construction program to develop permanent facilities. Due to budget limitations, the program had to combine better use of existing buildings and the creation of new facilities with flexible, efficient designs. The result was the 17,000-square-foot Site Two development project designed by an interdisciplinary team of young but experienced architects, engineers, and consultants.

Student lounge, handy to the sun-drenched hallway.

Flexible, Low-Cost, Energy-Saving Space

To reduce the project's cost, save space, and conserve energy, existing buildings were modified and linked by a new multi-use space unit with passive solar features. New offices and work stations were designed that can be removed or adjusted within hours to create more or different kinds of space. Within the 17,000-square-foot area are such college uses as laboratories, classrooms, faculty offices, and the college computer center. Lighting, heating, ventilation, and air conditioning are designed to accommodate multiple uses of the new facilities for a variety of rapidly changing functions. For instance, faculty offices can be converted in a matter of hours to space needed for a large meeting or seminar.

In sum, the Site Two design fulfills several goals common to many small educational institutions: with limited capital funding new space was produced that is adaptable to changing educational programs; the functional and visual qualities of the existing facility were improved through careful planning and placement of new space; and the operating cost of the new facility was minimized through energy-conservation measures.

In addition, this project demonstrates that passive solar concepts typically associated with residential design can be used in an institutional structure. Due to its modest size and flexible space, this particular building is well suited to these domestic scaled concepts. This approach in an educational setting is especially valuable because it increases acceptance for this type of energy-saving design by providing direct experience to young people about to enter and influence the world around them.

Energy-Conserving Features

Three major, passive, energy-conserving techniques, including passive solar heating, were used in conceptualizing this design. In the office spaces, south-facing clerestories with concrete mass walls below provide solar heat. Architectural scoops on the roofs bring passive heat into the north sides of buildings from overhead. The corridors are designed to be almost 100 percent solar heated: by tolerating a wider allowable comfort zone in circulation spaces, the temperature of the air is allowed to swing and work the thermal mass.

The second technique is daylighting to reduce electrical costs. Luminaires in the corridors and office areas are switched off automatically, by zone, when daylight provides sufficient lighting. The beige mass walls in the offices provide a good balance between heat absorption and light reflection into the room. And since daylight is introduced from overhead, there is an even distribution of lighting through the space without high contrast.

The third technique is natural ventilation. The corridors are designed with very large areas of operable glazing for ventilation. The building orientation provides maximum potential cross-ventilation from the prevailing wind. In addition, night insulation on most large glass areas helps conserve energy. Window overhangs are sized to reduce solar heat gain during the summer.

In addition, the building's mechanical system is designed to incorporate several energy-conserving features. For example, the fans have forward-curved blades and inlet vanes, a combination that provides excellent operating characteristics under partial loads. The economizer ventilative system is controlled by an enthalpic sensor in this very humid climate. Diffusers with a relatively wide volume range were incorporated to reduce the minimum volume of air delivered to the spaces, thus saving fan energy.

Programming the Form of Passive Energy

The functional programming and schematic planning of such a project are where the decisive energy opportunities must be recognized. Here corridors were conceived both as education spaces and as essential as a circulation and service spine to the multi-use spaces that primarily generated the need for building. But the diagonal linkage necessary within the existing campus plan did not discourage solar use just because the orientation was 45° off due south.

The single-loaded corridor can be an effective winter direct gain space in the morning. This is the heavy use time of day as well as the period when a building has cooled off the previous night and needs a fast warm up. Movable awnings and fixed shades protect the corridor windows in summer.

By placing the solar clerestory scoop facing due south over the roof, an interesting building geometry has been developed, and walls can best serve multiple uses since they are not pierced by windows. The clerestories provide the best orientation for day-long solar heating and daylighting, as well as seasonal control through calculated overhangs.

Designing to Expand Resources

For a community college as for any institution, resources are seldom enough to meet needs. But any activity that can reduce future financial obligations has the potential of design and physical expression. Solving such problems is a good beginning for architectural responsibility. Passive conditioning of spaces to avoid the perpetuity of fuel bills is perhaps easiest to visualize in a small institutional setting such as Site Two of Atlantic Community College. But the educational message can apply to any shape of building and any size institution, and the building opportunity to any responsible designer.

The simple directness of the original design drawings have been matched by the direct clarity of the completed construction. There was nothing tricky about the passive design concepts as abstract ideas, and there is nothing complex about the handsome building that resulted from this solution to the problem of limited resources.

A Drugstore In A Desert Climate

BUILDING

Trust Pharmacy
Grants, New Mexico

OWNER

Gene Carrica
Diamond "G" Trust
Grants, New Mexico

DESIGNER

Mazria/Schiff & Associates
Box 4883
Albuquerque, New Mexico 87196
Edward Mazria
Marc Schiff
Tom Cain, job captain
M. Steven Baker, mechanical engineering

BUILDER

Andrew Mirabel, contractor
Grants, New Mexico

CONSULTANT

August F. Mosimann
Engineering Associates, Inc.
1801 Lomas Boulevard, N. W.
Albuquerque, New Mexico 87104

STATUS

Occupied 1981

Estimated Building Energy Performance
(for the original designs)

BUILDING DATA

Heating degree days	5,600 DD/yr.
Cooling degree days	1,200 CDD/yr.
Floor area	3,420 sq. ft.
Building cost	$150,000
Heating	
Area of solar-heat collection glazing	550 sq. ft.
Solar heat	68 percent
Auxiliary heat pump	COP = 2.0
Cost of auxiliary heat	9¢/sq. ft./yr.
Cooling	
Two-stage evaporative coolers	
Auxiliary cooling needed	1,991 Btu/ sq. ft./yr.
Cost	3¢/sq. ft./yr.
Hot Water	
Need 3,600 gallons/yr. at 110° F. (incoming 55° F.)	
Auxiliary cost	$24/yr.

ANNUAL ENERGY SUMMARY

Nature's Contribution	
Heating	100×10^6 Btu
Cooling	No acceptable calculation
Hot water	1.6×10^6 Btu
Auxiliary Purchased	
Electricity	35.9×10^6 Btu or 9,702 Btu/sq. ft.

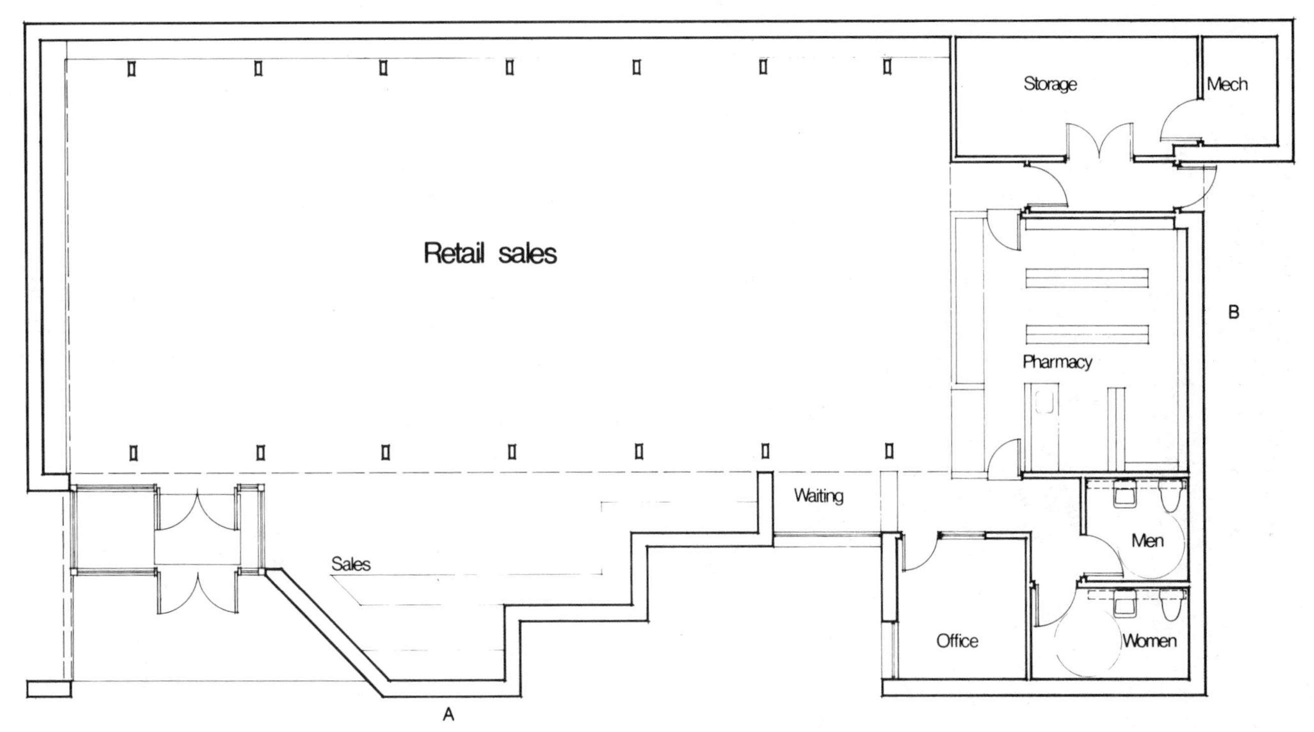

Plan of this building is simple, direct—and effective.

Generic Drugstore

The Trust Pharmacy is located along the commercial strip in the small, flat, windy community of Grants, New Mexico; a uranium boom town on the high desert plateau with severe winters and hot summers. The building is on the north side of a corner lot. This permits off-street parking and assures solar access. Its elongated east-west axis gives the building a long solar exposure.

This small commercial building is an example of integrating passive and hybrid heating and cooling strategies in a nonresidential structure. The principal passive environmental elements require no personal interaction, so the building can be enjoyed as a quiet, stable space with no moving parts. To free all wall surfaces for merchandizing, the passive features are overhead.

Beyond the economy of a passive thermal and luminous environment is the architectural quality of this small, simple, direct building. Neither fussy nor self-conscious, its memorable character comes from the glowing louvered ceiling that admits heat and daylight. This handsome shop would be the pride of any community.

Daylight From Above

Because this pharmacy is primarily a day-use facility, daylight is an obvious passive design strategy. Winter solar heat collection has been integrated with year-round natural lighting. To make the most commercial use of floor and wall surfaces and to minimize security risk, the designers chose overhead glazing instead of low operable windows; the windows in the sawtooth roof provide approximately 90 percent of the lighting. Long, horizontal windows on the roof are also easier to shade seasonally with fixed overhangs. Interior louvres in the ceiling sawtooth openings diffuse solar heat gain over the thermal storage surfaces and provide for an omnidirectional, well-balanced, glare-free light source during daylight hours.

Post-and-beam system of interior is carried outside to arch beside entrance.

Heating and Cooling

The Trust Pharmacy is heated primarily by a direct gain, passive solar system. The building was built of solid masonry to provide a sufficiently large surface mass in what is basically an open plan. The large interior mass would store solar heat gain admitted during winter days to buffer against severe temperature swings. In summer, the masonry would delay the impact of the sun and high outdoor temperatures until evening when the building is not in use.

Because natural ventilation is not possible in this windy and dusty region, a residential-size, two-stage evaporative cooler system was proposed to filter the air and cool the interior at night. It would provide cool interior surfaces for the following day by operating primarily during off-peak hours. In summer, new vegetation would shade the west facade to diminish solar impact on that wall. Calculated overhangs would shade the south-exposed glass while maintaining adequate light levels inside.

A Fine Design, But Too Expensive

Thus the pharmacy was conceived as a toplit warehouse in which every architectural surface would participate in passively conditioning the wide-open commercial space. Two full-length windows on the entrance elevation would allow views of the outdoors and admit direct sunlight. It was a fine design that appeared to satisfy every need. The calculated thermal performance indicated that the building would be highly efficient, with a low dependence on off-site energy. Unfortunately, the building as designed could not be built within budget.

Cutting Costs and Improving Energy Performance

Fortunately, it was possible to reconsider the construction and the mechanical systems to lower substantially the construction cost and yet maintain the passive and hybrid concepts. The original design used heavy commercial construction standards. It was to be a brick building with a metal roof, aluminum soffits, and a complex cooling system. This "Cadillac" package was revised to a "Chevrolet" approach by using construction typical of custom-built homes. As built, the exterior is concrete block, fluted up to 8 feet and then split face above. The inside is furred drywall. In contrast, the original design proposed a cavity wall with brick outside, an 8-inch grout-filled concrete block inside, with 4-inch polystyrene insulation board between. The original heavy-duty commercial door and window frames were replaced with high-quality residential standard frames. The roofing of the revised building is heavy-duty asphalt shingles. Thus the building was completed at a cost of $15 per square foot less than originally designed, yet with improved energy performance.

Sawtooth roof provides approximately 90 percent of the lighting.

Saving With Mechanical Equipment

Another dramatic economy came from changing the mechanical design from a heat pump and two-stage evaporative cooler system to a gas furnace and a direct evaporative cooler. The change saved 50 percent on the equipment cost and resulted in a lower operating cost.

Since the pharmacy is not open on summer evenings, the evaporative cooler can be used then, when its performance is best. By lowering the temperature of the interior thermal mass at night, the interior is cooled for the next day. This greatly simplified the requirement for cooling equipment. Similarly, a gas furnace for winter auxiliary heating is much simpler to install and operate than an electric heat pump and duct system.

Using Plasterboard for Thermal Mass

Perhaps the most interesting change between the preliminary design and the Trust Pharmacy as

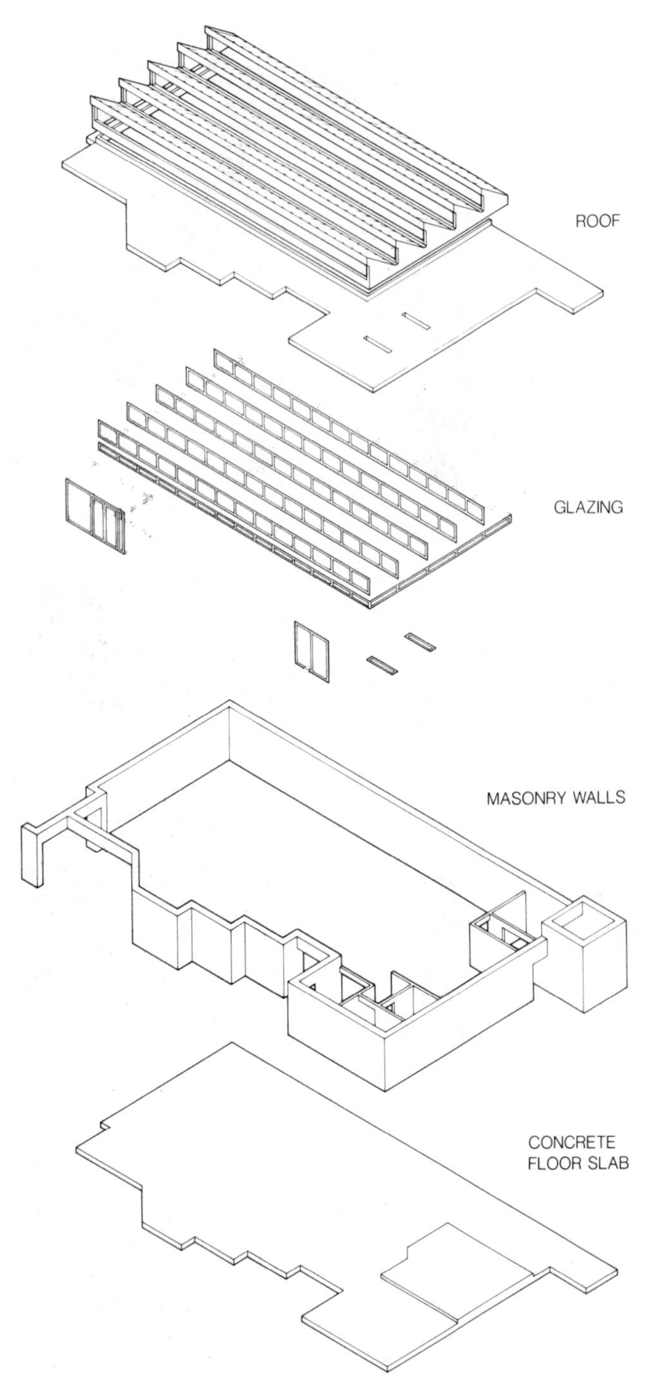

The passive solar components are shown in this exploded isometric drawing. From top, the sawtooth roof that permits sunlight to flood store; the long, horizontal windows; the one-inch drywall plasterboard walls, for thermal storage; and the floor, a concrete slab topped with quarry tile.

built is the interior materials. In a passively heated and cooled structure, the thermal capacity and exposure of the interior mass are critical for maintaining comfortable conditions. Here, instead of interior exposed concrete block, the 4-inch insulation furring is covered with 1-inch drywall plaster board. This special material is normally used to achieve the substantial fire ratings required for partitions in public buildings. In the Trust Pharmacy, it is used for its thermal storage capacity—double that of ordinary drywall plasterboard. Each sheet weighs over 200 pounds, which presented a construction challenge to the builder. Together with the ⅝-inch drywall plasterboard on the slopes of the ceiling between the clerestory windows, and quarry tile over the concrete floor slab, there is a ratio of 12 to 1 for the area of useful thermal mass to the energy collection glass. Thus, the masonry walls were not needed for thermal storage. In this particular day-use building, slightly larger diurnal temperature swings were acceptable than would be for a residence.

Measuring the Mass of the Merchandise

It was known that the thermal mass of the pharmacy's merchandise could also contribute to the thermal stability of the interior. There was, however, no easy way to estimate the thermal quantities involved. Thus, they were ignored in the preliminary performance calculations. The actual performance of the building has been better than calculated, presumably due to the uncalculated thermal mass of the contents of the sales shelves.

Appearances and Performance

The dramatic visible impact of a solar design that is not an "add-on" approach but is integrated can best be seen on the light-filled inside. The roof construction is articulated on the interior as a structure floating free over the walls with the ceiling separated by a continuous clerestory of glass at

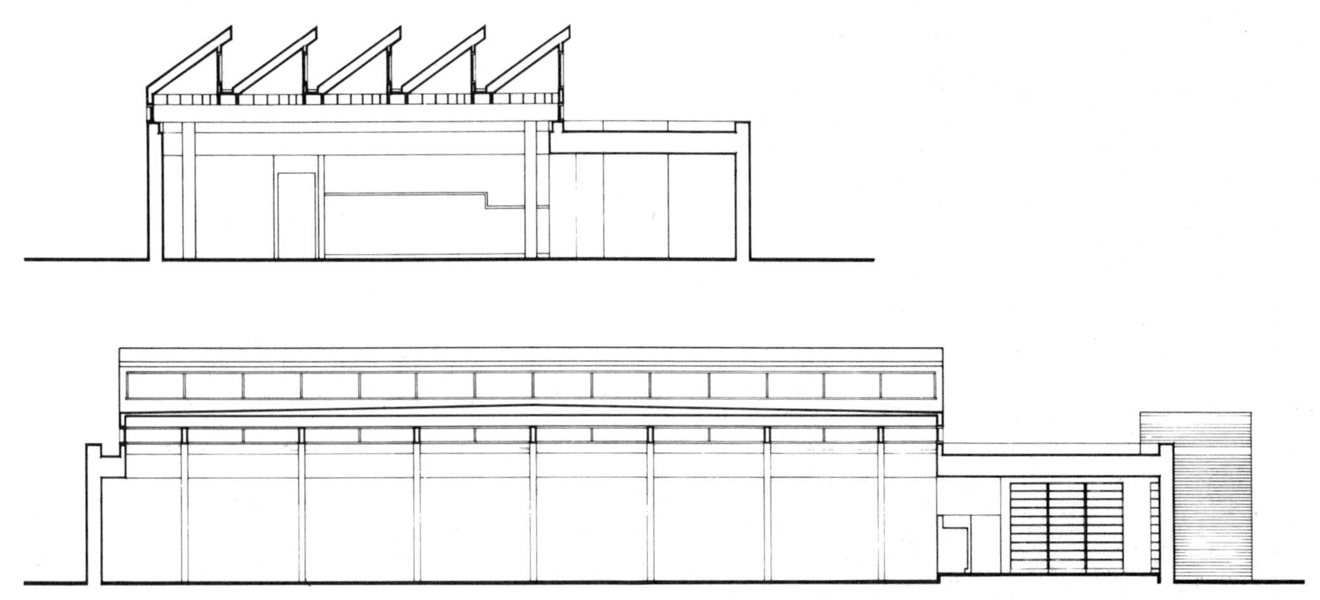

Section A (top) and B, as shown on page 245, illustrate prominence of sawtooth windows.

the top of the wall. The exposed wood post-and-beam system stands clear of the walls and visibly organizes the mixture of merchandise typical of an American drugstore. The continuous flat ceiling alternates bands of white louvres with the sparkle of fluorescent fixtures used as auxiliary lighting and mounted on the bottom of the solid structure. The wood louvres are spaced unevenly to conceal the sawtooth windows and to bounce daylight into a non-directional matrix that fills the store. Thus, natural light and solar heat filter from above, providing a sense of order and space to make an otherwise commercial scene seem to be an almost inspired space to buy.

The overall operating experience of the Trust Pharmacy has confirmed the prudence of thermal design decisions at every level. The proof is in the utility bills, which for December 1981 and January 1982—among the coldest months of recent winters—were only $22 monthly. The visible evidence of this energy economy is seen in the distinctive sawtooth skyline outside, and the rich ceiling inside that handsomely integrates natural heating and lighting and flatters the processes of commerce.

Office Park With Innovative Cooling

BUILDING

Princeton Professional Park
2303 Whitehorse — Mercerville Road
Mercerville, New Jersey 08619

DESIGNERS

Short and Ford, architects
RD 4, Box 864
Mapleton Road
Princeton, New Jersey 08540
Harrison Fraker, architects
575 Ewing Street
Princeton, New Jersey 08540

CONTRIBUTORS

Princeton Energy Group
575 Ewing Street
Princeton, New Jersey 08540

BUILDER

The Karnell Group
202 12 Street
Piscataway, New Jersey 08854

STATUS

Occupied Summer 1983

Estimated Building Energy Performance

BUILDING DATA

Heating degree days	4,908 DD/yr.
Cooling degree days	968 CDD/yr.
Total floor area	64,900 sq. ft.
Volume of conditioned space	849,060 cubic ft.
Building cost	$3,009,000
Cost of special energy features	$394,789
Total building UA daytime	27,134 Btu/°F./hr.
Total building UA night	19,028 Btu/°F./hr.
Heating	
Net area of solar-heat collection glazings	14,096 sq. ft.
Solar heat	60 percent
Auxiliary heat (heat pumps)	13,700 Btu/sq. ft./yr.
Cost of auxiliary heat	14¢ sq. ft./yr.
Cooling	
Net roof heat dissipation	
Auxiliary cooling (heat pump)	4,380 Btu/sq. ft./yr.
Cost of auxiliary cooling	6¢ sq. ft./yr.
Lighting	
Daylighting contribution	90 percent
Need 9.6 × 10^6 Kwh/sq. ft./yr.	2,840 Btu/sq. ft./yr.
Cost of electric lighting	50¢/sq. ft./yr.
HVAC Fans	
Need 2.79 Kwh/sq. ft./yr.	9,510 Btu/sq. ft./yr.

ANNUAL ENERGY SUMMARY

Nature's Contribution	
Heating	1,616 × 10^6 Btu
Cooling	251 × 10^6 Btu
Lighting	2,090 × 10^6 Btu
Auxiliary Purchased	
Electricity	3,430 × 10^6 Btu/yr. or 53,600 Btu/sq. ft.

A COMPARATIVE ENERGY ANALYSIS

	Base Case Typical Speculative Office Building	Princeton Professional Park Solar Building
Heating		
End use Btu—Btu/sf/yr	48,200 (30%)[1]	13,700 (25%)[2]
$/sq. ft.	.48	.14
% of total building energy cost	23	20
Cooling		
End use Btu—Btu/sf/yr	37,400 (23%)	4,380 (8%)[3]
$/sq. ft.	.52	.06
% of total building energy cost	25	9
Lighting[4]		
End use Btu—Btu/sf/yr	35,500 (22%)[4]	2,840 (5%)
$/sq. ft.	.50	.04
% of total building energy cost	24	6
Miscellaneous		
DHW[5]		
End use Btu—Btu/sf/yr	11,600 (7%)	5,800 (11%)
$/sq. ft.	.16	.08
% of total building energy cost	8	11
HVAC Fans[6]		
End use Btu—Btu/sf/yr	11,200 (7%)	9,510 (18%)
$/sq. ft.	.16	.13
% of total building energy cost	8	19
Equipment[7]		
End use Btu—Btu/sf/yr	17,700 (11%)	17,730
$/sq. ft.	.25	.25
% of total building energy cost	12	36

[1] Assumes 75 percent efficiency at a rate of $10/10^6 Btu

[2] 218.6 × 10^6 Btu/16,000 sf at a rate of $10

[3] (1/2 May, 1/6 June, 1/3 July, 1/4 Aug., 1/6 Sep.) at a rate of $14/10^6 Btu

[4] at 4 watts/sf × 10 hr/day × 260 days/year at the rate of $14/10^6 Btu

[5] one person/150 sf @ 10 gal/day × 260 days × 8.33 Btu/gal/°F. × 80°F. at a rate of $14/10^6 Btu

[6] one unit/1,000 sf @ 1/2 HP (500W) × 365 days × 18 hr × 3.41 Btu/HR-W at a rate of $14/10^6 Btu

[7] at 2 Watts/sf × 10 hr/day × 260 days/year at a rate of $14/10^6 Btu

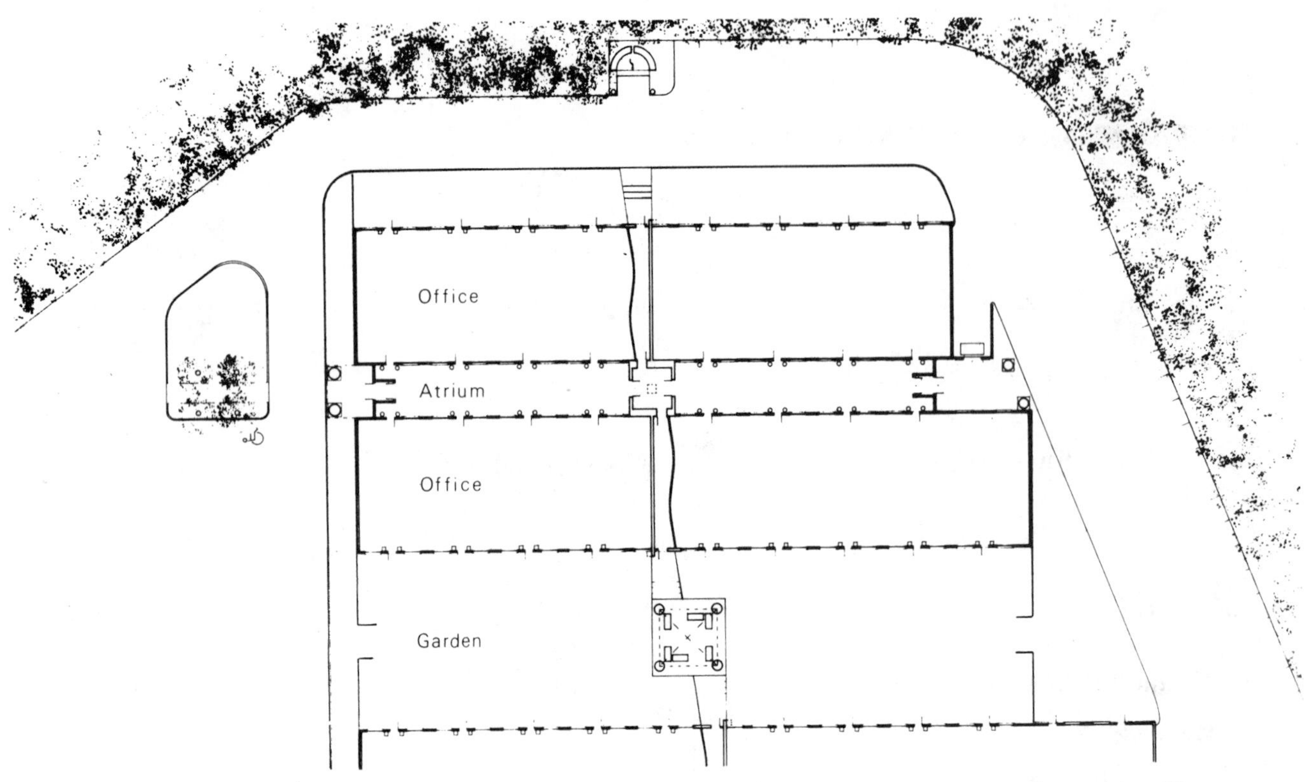

Floor plan of the northernmost of the three pavilions, showing the two single-storied, 45-foot-wide office structures linked by an atrium roofed with clear double acrylic. Parking is at left.

Although the site and the building requirements for the Princeton Professional Park were typical, the solution was not. This interesting design of offices for the speculative developer market was the result of a joint venture by two small architectural firms. The collaboration provided a unique solution to the cooling and daylighting needs that dominate energy uses in this well-known building type. Beyond the imaginative integration of environmental needs within the fabric of the structure a particular architectural distinction was achieved—all within the economic constraints of a tight budget.

Commercial Spaces in an Arcadian Setting

The generous ten-acre site in Princeton Township is located along a primary bus route. Its wooded character and gentle south-facing slope encouraged the designers to continue the arcadian setting for this new commercial use. In addition, the potential for passive heating and cooling could be enhanced by the development of a garden surrounded by a band of trees.

The elementary organization of the site provides an easy understanding for first-time visitors and workers. The visitor parkway is on the west with a direct link through a wooded area to the major artery. Staff and employee parking is separated on the east side of the building and has its own access road.

The complex should provide a 64,000-square-foot professional office complex for doctors, dentists, lawyers, architects, etc., with a minimum rentable office module of 1,100 square feet. These needs resulted in an economical wood construction with prefabricated 9-foot-wide plywood-faced panels and 36-inch-deep wood trusses.

The building is a series of three pavilions of

different lengths each with a lineal atrium as a broad circulation access. Building pavilions are separated by walled gardens. An open lawn and pond at the south end of the property are connected by a transverse walk, with undulating walls, that slices through the three pavilion buildings. Fountains and aedicular kiosks mark the nodes of this path that links the gardens. Its north-south cross axis counters the workaday vectors between parking lots and work spaces.

Each pavilion consists of two single-story, 45-foot-wide structures. Their shed roofs slope upward toward the high lineal atrium that is roofed with clear double acrylic to form an internal landscaped sunspace.

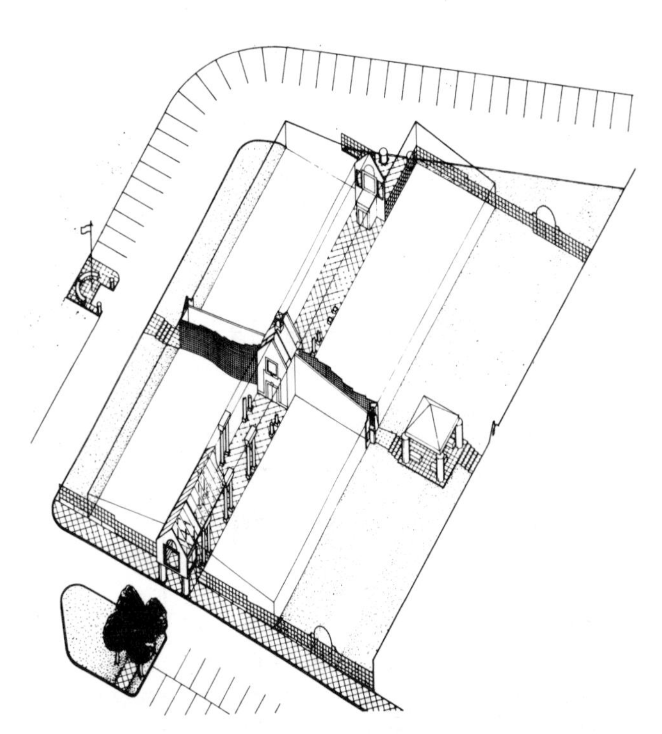

View from above shows how roofs of office sections slope upward over the atrium.

From Energy Analyses to Architectural Form

The design team examined the annual energy use of a typical speculative office building in the vicinity of Trenton, New Jersey, as a prelude to the hypothesis of architectural form. It was apparent that the heating load did not constitute the major portion of the energy consumed by this building type. In fact, on a cold, clear winter day, the daytime internal heat gains practically equalled the losses.

This energy analysis showed that typically heating, cooling, and lighting were roughly equal in annual cost, and that reduction in heating load should not be a primary concern. Rather, the greatest benefits could result from reducing energy used to light the building. This would reduce electrical demand and reduce the cooling needs as well.

A thermal and functional analysis also showed that solar heating energy collection should not occur primarily within the occupied space. An atrium concept made it possible to collect heat outside the office space, and to store it in a horizontal rock bed under the office floor for use during the night when internal gains are not present.

Yet another concern was to minimize the mechanical cooling needs of the building. The extensive use of daylight to replace fluorescent lights represented a significant reduction with which to begin. Further reduction was accomplished by spraying the roof with water during the day. Evaporation cools the roof directly and can eliminate extensive heat gain from the sun.

In addition to this immediate heat reduction, the roof spray could be operated at night, when a combination of evaporation and radiation of heat to the sky would reduce the roof temperature below the wet bulb temperature. Under good conditions of some wind and a relatively clear sky, the cooling effect could be as low as the dew point temperature. This cooling effect could be exploited by circulating air from a narrow plenum directly under the sheet metal roof through a rock bed. Weather data plotted for Newark, July 22–31, 1975, showed that the nighttime dew point temperature is almost always below 65 ° F. Thus, the system should be able to almost completely eliminate the need for mechanical cooling throughout the year.

The use of continuous rock beds for thermal storage under the floor slabs integrates this cooling within the building fabric. The same thermal stor-

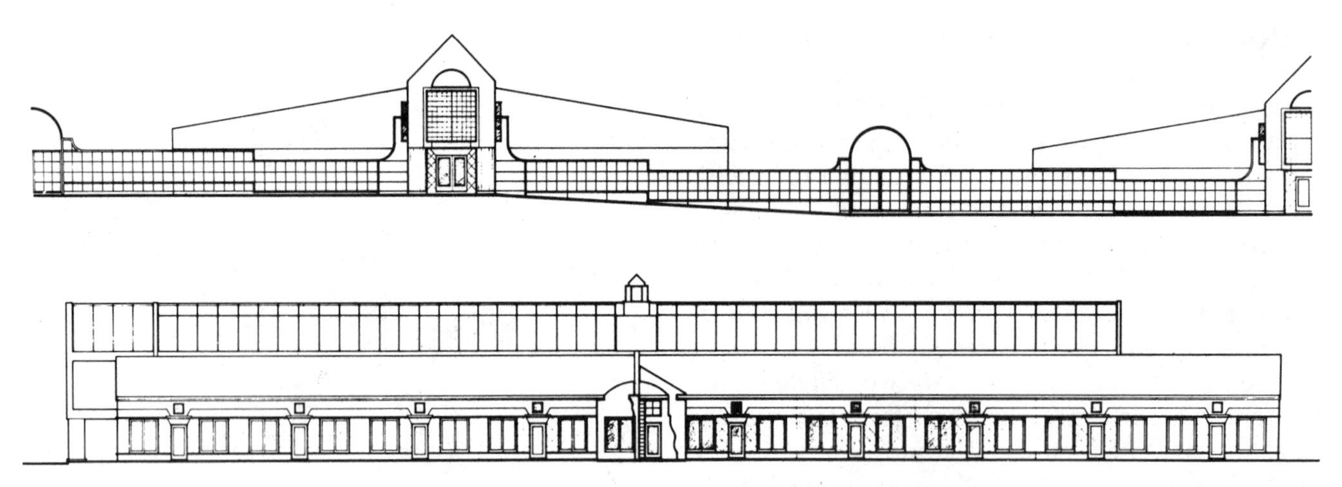

Entry facade, above, and south elevation.

age can be used for solar-produced heat during winter, as well as a means of balancing thermal conditions of the interior without throwing away a potential thermal advantage. By being located directly under the space they serve, any thermal leaks from these rock beds are not really lost. The construction of rock beds under the slabs is an extension of typical construction practice and is economical.

Calculating the Energy Performance

After a schematic solution to the building was proposed, a detailed comparison between the design and a typical "base case" building similar in dimensions was calculated. Since the developer/client had built similar projects in the area, it was possible to use a realistic data base. The energy economy in Princeton Professional Park is achieved by the following means:

Heating

- Indirect solar energy collection in atrium with thermal storage in horizontal rock bed and slab for nighttime loads.
- Modest direct solar gain through windows with thermal storage in floor slab.

Daylighting

- Atrium distributes natural light through clerestory windows to offices which have two options for interior planning: 1) either a typical ceiling with a light shelf and translucent panels, or 2) an open clerestory with landscape office planning.
- Stepped electric lighting controls adjust to day lighting levels automatically or by hand.

Cooling

- Natural ventilation is induced through the atrium by thermal and wind pressure when outside conditions are suitable.
- Load reduction during daylight hours is achieved by evaporation from spraying the roof.
- Cooling potential is stored in the rock bed at night from air cooled by circulating air under the metal roof while it is sprayed. Cooling is achieved by evaporation and nightsky radiation. During the day the return air supplied to the space is precooled in the rock bed. If this meets comfort requirements, no refrigerated cooling is necessary.

Operating costs for heating, cooling, and daylight energy in the base case building is 2.14 per square foot per year. Load reductions achieved in the design result in predicted energy costs of $.54 per square foot per year. This represents an 80 percent savings in total energy cost at an add-on cost of $3 per square foot and allows a five-year payback.

A simple first-year return on investment analysis yields a 23 percent return on investment. This analysis does not account for fuel inflation, possible tax credits, property tax reductions, or sales tax rebates, all of which would make the project an even more attractive investment in alternative methods of evaluation. Since the results of this sample analysis were convincing and fully satisfied the client's requirements, more elaborate analyses were performed and the building was constructed. Results to date have confirmed the economy of performance anticipated in the design.

This perspective shows relation of office space, left, with lofty atrium.

Architecture From an Energy Filter

Although by definition a rental office must be an anonymous place, Princeton Professional Park has a character that is far from forgettable. "The principle of using the building form and envelope as a dynamic filter between climate and the natural provision of comfort" generated the architectural form. The atriums as the mediators in filtering outdoor environment for interior conditions have also become the most visible part of the architectural expression. The high gable ends of the atriums are visually emphasized on the exterior with a pedimental portico treatment that forms a gateway into the axial circulation with its daylit greenhouse atmosphere. Ventilation grilles and exhaust cupolas are prominently and formally visible on axes. Trellises soften and define.

The client allowed only $37 per square foot for construction—the low end of building budgets. Added costs for energy had to satisfy a five-year payback. This stringent budget has still allowed a high performance building to be completed that attracts tenants by its energy economics as well as its pleasant and memorable character.

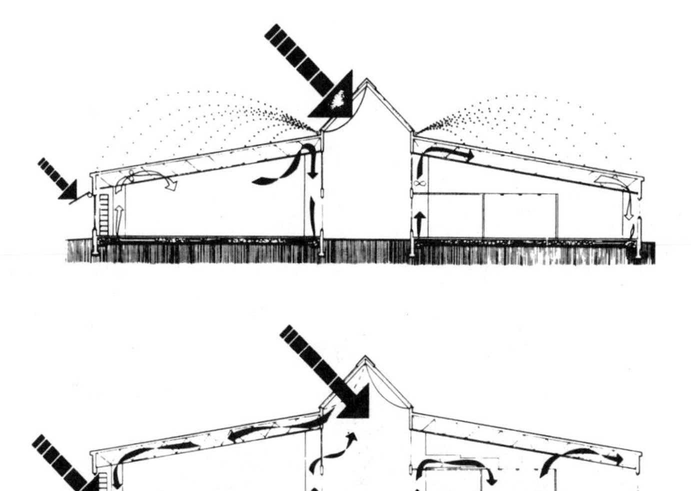

In top drawing, left, cooling during daytime is achieved by spraying roof. At right, roof is sprayed at night. Air circulating under it is cooled, then circulates through and cooks the rock bed. In bottom drawing, left, heat during the day is stored in rock bed. At right, stored heat is circulated at night.

100 PASEO de SAN ANTO

A Humanized Government Office Building

BUILDING

State Office Building
San Jose, California

DESIGNER

ELS Design Group/Sol Arc
2040 Addison Street
Berkeley, California 94704

BUILDER

E.P. Lathrop Company
Emeryville, California 94608

STATUS

Occupied June 1983

CONTRIBUTORS

Berkeley Solar Group (computer analyses)
3026 Shattuck Avenue
Berkeley, California 94705

Guttman/MacRitchie (mechanical engineer)
1300 Sutter Street
San Francisco, California 94109

Engineering Enterprise (electrical engineer)
620 Bancroft Way
Berkeley, California 94710

Estimated Building Energy Performance

BUILDING DATA

Heating degree days	2,500 DD/yr.
Cooling degree days	250 CDD/yr.
Volume of conditioned space	990,000 cubic ft.
Total floor area	130,000 sq. ft.
Building cost	$11,500,000
Cost of energy features	$400,000
Total building UA	26,660 Btu/°F./hr.
Heating	
Net area, solar collection glazings	3,500 sq. ft.
Energy required, gas	13,000 Btu/sq. ft./yr.
Cooling	
Energy required, electricity	4,000 Btu/sq. ft./yr.
Hot Water	
Need 226,000 gallons/year at 110°F. (incoming 55° F.)	
Lighting	
Need 2.05 Kwh/sq. ft./yr.	7,000 Btu/sq. ft./yr.
HVAC Fans	
Need 1.025 Kwh/sq. ft./yr.	3,500 Btu/sq. ft./yr.

ANNUAL ENERGY SUMMARY

Nature's Contribution	
Cooling	6,000 Btu/sq. ft.
Hot water	900 Btu/sq. ft.
Lighting	11,000 Btu/sq. ft.
HVAC	3,500 Btu/sq. ft.
Auxiliary Purchased	
Electricity and gas	3,100 × 10^6 Btu or 27,900 Btu/sq. ft.

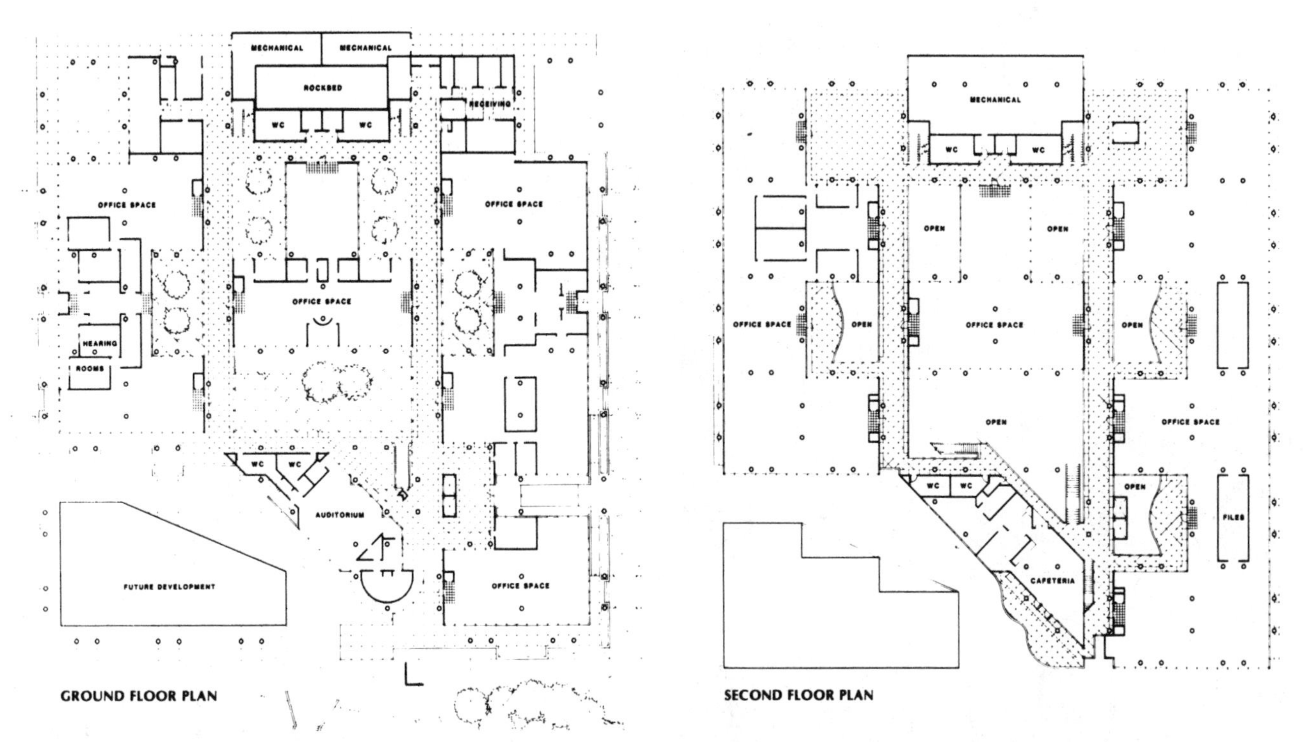

Ground and second floor plan emphasize designer's determination to create a checkerboard of courtyards to organize twenty-two agencies in a single, three-story building.

This government office building, with more than 130,000 square feet of floor space, makes optimum use of conservation and passive energy techniques to reduce both energy and operating costs. But the principal value of this design is that the energy-saving features, such as daylighting, small work spaces close to windows, and landscaped courtyards, also create a more humanized environment. Furthermore, as a lowrise office building in the midst of San Jose's downtown parking lots, it contributes urbanity to an intense mixed-use redevelopment of the immediate eight-block area. The attractive courtyard scheme is in keeping with the community's historic flavor; arcades along the building perimeter extend the protected pedestrian environments of the city. Yet, the approach to amenity, both inside and out, is simple and economical. As a kind of shopping center for 22 different government services or agencies, this unassuming building demonstrates that a design can serve an energy-conserving role as well as a programmatic function.

The San Jose Climate

The mild climate of San Jose is particularly amenable to passive energy applications and conservation strategies. The average summer high temperatures are in the low 80s, and wintertime lows average in the mid 40s. On the hottest summer days, temperatures rise into the 90s, but fall into the 50s or 60s at night. Typically on these days, the humidity is unusually low. Such conditions are ideal for cooling by night heat removal from the building structure, and the use of a chargeable rock bed for cool storage.

An average of 83 percent of possible sunshine is typical in the summer, but solar gains can be effectively controlled by proper shading. In winter, an average of only 45 percent of possible sunshine is available. The record low in San Jose is 20° F. The ASHRAE 2½ percent cooling design tempera-

Entrance provides easy access to all offices.

ture is 88 ° F. and the 1 percent heating design temperature is 34 ° F. There is an average of 2,438 degree days of heating required per year of which 488 occur in January. These climatic statistics are clear indications that there are major heating and cooling problems in large buildings and also that their range invites major passive intervention. They prompted the computer examination of various strategies especially responsive to such weather patterns.

Computer Model Sources and Investigations

The energy simulations used in the computerized thermal analysis were based on the California Thermal Zone #4 Weather Tape prepared for the California Energy Commission. Although the hourly weather tape purports to represent an "average" year, the average winter temperatures on the tape were significantly higher than the long-term averages for San Jose, and an allowance was made for these differences.

Energy consumption for heating and cooling was estimated using a computer simulation model developed by the Berkeley Solar Group (BSG). The building model consists of NBSGLD, a thermal model which calculates the heating and cooling loads for each type of space and exposure; ZONE, an aggregation program that assembles the results of several load runs into a loads file for each fan zone; and SYSTEM, a program that simulates the performance of a variable volume air conditioning system serving up to 10 fan zones.

NBSGLD, the space load program used in the simulations, is an extensively modified version of NBSLD, the load model developed at the National Bureau of Standards by Kusuda and others. BSG has, in addition to correcting errors in the original program, added several features that improve the modeling of energy-conserving office buildings. One feature particularly useful in this project is a night ventilation subroutine that models the effects of ventilating the space at night with evaporatively cooled outside air. Other BSG modifications allow the modeling of interior surfaces such as structural members and furniture; and the modeling of hung ceiling return air plenums that contain significant thermal mass and through which return air and heat from lights are exhausted. An additional subroutine was developed for this office building as a project that allows a crude simulation of the effect of natural daylighting by varying the artificial lighting according to the amount of solar radiation transmitted through the windows.

Simulation runs using the space load model were made for each significantly different load condition in the building. Runs were made on a standard 1,000-square-foot module for 29 different exposure and floor combinations. There were more load conditions in this building than a "normal" building because of the additional exposures in the courtyards.

The ZONE program was used to aggregate the loads from individual spaces into a load file for each of six fan zones. ZONE scales the hourly load file produced by the space load program for each exposure according to the floor area of that exposure in the fan zone. The resulting zone load file includes total air flow, supply and return temperatures, reheat energy if required, and electricity consumption for lights and equipment.

The SYSTEM program models the mechanical system. Equipment models and part-load performance curves from the CAL-ERDA computer program were used for fans, chillers, and boilers. The rock bed subroutine is a standard multinode model based on a program developed at California Polytechnic State University, San Luis Obispo, by Philip Niles and used successfully in the design of a large rock bed in Fresno. The evaporative air washers are modeled using a constant evaporative effectiveness. The computer output of the system model, in addition to detailed reports on the operation of the mechanical equipment, provides the estimated total annual electricity and gas consumption for the building, and the energy costs based on current fuel, electricity, and peak demand charges.

Note that the domestic hot water system and the cafeteria and kitchen needs were not modeled and their energy consumption is not included in the annual totals.

One of the landscaped courtyards that enhances the quality of the working environment.

Individual Zone Loads—Cooling/Heating

Table 1 shows the peak cooling loads derived from the computer simulations. The major differences between north and west and between east and south zones are attributable mainly to the effects of the sun-shading. The peaks on the north and west are due to late afternoon direct solar gain. They can be reduced by redesign of the shading on these facades. The relatively high peaks in the courtyard zones occur in March, when the overhead shading curtain is assumed to be open. Providing this curtain is operating by a mechanism capable of closing it on any warm day, these peak loads can be expected to decrease markedly. The monitor skylights on the top floor are responsible for the large loads in those interior zones.

Table 1 also shows the peak heating loads. These loads are all early morning start-up loads occurring on Mondays after cold, cloudy weekends. The zone-to-zone differences are attributable mainly to variations in the amount of direct solar gain received and retained by the mass in each zone.

System Cooling Loads

For the recommended system configuration, the peak system cooling load is about 133 tons. This peak occurs on a very hot day (99° F. maxi-

Table 1

PEAK COOLING LOADS — Btu hr./sq. ft.

					Courtyard				
	N	S	E	W	N	S	E	W	Internal
TOP	28.1	19.2	19.2	30.8	19.1	30.5	19.1	35.7	31.1
MID	28.5	19.6	19.6	31.0	19.5	25.8	19.5	30.5	9.3
BOTTOM	26.8	—	26.1	32.4	25.9	30.5	—	26.0	9.3

PEAK HEATING LOADS — Btu hr./sq. ft.

					Courtyard				
	N	S	E	W	N	S	E	W	Internal
TOP	19.8	17.0	18.7	19.4	19.8	16.8	19.8	18.0	21.4
MID	19.6	16.2	18.5	18.9	19.6	18.1	19.6	19.1	7.7
BOTTOM	23.6	—	22.4	22.9	23.6	23.5	—	23.6	5.5

mum) following a warm night (58° F. minimum). These outdoor air temperatures limit the capacity of both the structure of the building and the rock bed to flush out heat at night. The relatively mild San Jose climate produces very few "peak" loads. Even with the ability of the rock bed to meet most of the partial loads, the low annual net chiller COP reflects the amount of part-load chiller operation required.

Night Ventilation

The design of the night ventilation/mass system depends on finding both a functional night air volume consistent with the air-handling system's capacity, and a practical control strategy that will yield both net energy and energy cost savings.

The building's air-handling system was designed to provide a minimum volume of .7 cfm (3.5 air changes per hour) per square foot of office space—potentially the lowest controllable supply volume for the system. Previous studies on a similar building in the more extreme Sacramento climate indicated an "optimum" ventilation volume of between 3 and 4 changes per hour, consistent with the 3.5 air change minimum volume for this system. In those simulations the night ventilation system was run when the night temperature difference between indoor and evaporatively cooled outdoor air was greater than 4° F. on nights following days whose average temperature was above 70° F.

Using this same configuration and control, the system model indicated that with the night ventilation the building's annual energy cost is reduced slightly due to reduced demand charges; however, its overall energy consumption is increased by 4 percent as a result of the increased electrical consumption required to run the night ventilation fans. If the controls are changed to run the system at night temperature differences greater than 6° F. on days averaging 76° F. or greater, the fans run substantially fewer hours during the spring and fall. This results in about $800 per year savings, slightly more than 2½ percent of the building's energy cost, and does not significantly affect peak loads. Although the annual energy consumption is still slightly higher by ¾ percent, the peak cooling load is reduced by 15 percent, and 30 percent of the total chiller energy consumption is transformed to off-peak fan power consumption, producing the $800 per year savings on demand charges. Additional studies of various control strategies resulted in net annual savings in energy consumption.

The Role of the Rock Bed

The thermal storage rock bed is used to cool the building's return air stream. Heat absorbed in this manner during the day is removed at night by blowing cool outdoor air through the bed. The three major objectives in using such a rock bed are: to decrease the peak demand for chiller output, thereby reducing the peak demand for electricity; to reduce the required peak output and the first cost of the chiller; and to reduce the annual energy consumption of the building's cooling system.

Recommendations from the Initial Modeling

As modeled during the investigative phase, the building was estimated to consume approximately 26,300 Btu (net) per square foot annually. This consumption is less than 60 percent of that permitted under the energy budget provisions of the California Title 24 Non-Residential Standards. As part of the energy analysis, the individual contributions of the major subsystems were determined. These studies, starting from a base case with no natural-artificial lighting integration, no night ventilation, and no rock bed, indicated that the control of the lighting produced the greatest annual energy savings. The next greatest saving was produced by the rock bed, followed by the night ventilation and the use of two chillers. The estimated annual energy cost savings were $6,300, $2,500, $1,000 and $300 respectively. Thus, the provision of artificial-natural lighting controls, night ventilation, and a rock bed were recommended, and became a part of the design as built.

The configuration of the building's energy systems reflects three major goals in minimizing its energy requirements and costs: the reduction and shifting of loads placed on the systems; the utilization of the passive cooling potential of the building's structure; and the addition of thermal mass to augment the effects of the structure.

Loads placed on the energy systems were reduced through a careful integration of artificial and natural light, the reduction of skin loads, the ducting of return air through light fixtures, and the removal of direct summer solar gain through extensive sun shading. The installed artificial lighting was designed for a peak power input of less than 1.9 watts per square foot average. Employing a switching strategy that turns these lights off as the natural light level increases can drop the average year-long consumption to as low as 1.25 watts per square foot. Skin loads were reduced through the use of R-20 roof insulation and fully weather-stripped windows and doors. Analysis showed the use of double-glazed windows to save about $1,500 per year on the building's annual energy cost. This was not sufficient, however, to justify the increased cost of changing from single to double glass.

Prologues to Successful Design

Much of this description comes from the designers and consultants programming and preliminary design phases of this important office building design. It details parts of a building which if not always invisible, are usually taken for granted. It documents how these early design probes and prognostications can define the life history, especially the comfort potentials and thus the energy price for a large building investment.

Within those anticipatory activities the accretion of design experience with large buildings is evident. Particularly in California, the state support of energy studies and standards has been coupled with the state practice of building new structures that embrace and extend those energy concerns. Here the simulation of computer studies has benefitted from extended previous experience. But the reality of how energy design has influenced the architecture has to no small degree become invisible. There are no obvious clues in the completed building that dramatize how climate, energy, and life cost concerns dominated the formulation of the design.

A Horizontal Solar Skyscraper

BUILDING

Department of Justice Office Complex
Sacramento, California

DESIGNER

Marquis Associates
Architecture/Planning/Interior Design
243 Vallejo Street
San Francisco, California 94111

STATUS

Occupied 1982

BUILDER, PHASE I

RG Fisher Co.
Box 8026
Fresno, California 93727

BUILDER, PHASE II

Continental Heller Corp.
Box 2551
Sacramento, California 95812

Estimated Building Energy Performance

BUILDING DATA

Heating degree days	2,782 DD/yr.
Cooling degree days	1,640 CDD/yr.
Total floor area	350,000 sq. ft.
Volume of conditioned space	2,670,000 cubic ft.
Building cost	$28,000,000
Savings in cost of special energy features over conventional office construction	$280,000
Heating	
Internal heat gain from people; heat recovery from computers, chillers, equipment, and lights	
Auxiliary heating	None
Cooling	
Shading of south glass	
Night air flushing—cooling of thermal mass	
Underground chilled water storage to flatten cooling load	
Auxiliary, electric	40,000 Btu/ sq. ft./yr.
Cost of auxiliary cooling	$4.68/sq. ft./yr.
Hot Water	
Need 520,000 gallons of 120° F. (incoming 60° F.)	
Solar assisted, electricity cost	$2,600/yr.
Lighting	
Recessed fluorescent, one 80 watt plus one 40 watt fluorescent task/100 sq. ft.	
Cost of 2 Kwh/sq. ft./yr. (6,826 Btu)	8¢/sq. ft./yr.
Blowers	
Average 1.2 Kwh/sq. ft./yr.	4,000 Btu/ sq. ft./yr.
Elevators/Escalators	
Average .15 Kwh/sq. ft./yr.	500 Btu/ sq. ft./yr.

ANNUAL ENERGY SUMMARY

Nature's Contribution	
Heating	Cannot be
Cooling	isolated
Hot Water	174.4×10^6 Btu
Lighting	Not available
Auxiliary Purchased	
Total	$1,488 \times 10^6$ Btu or 40,000 Btu/ sq. ft.

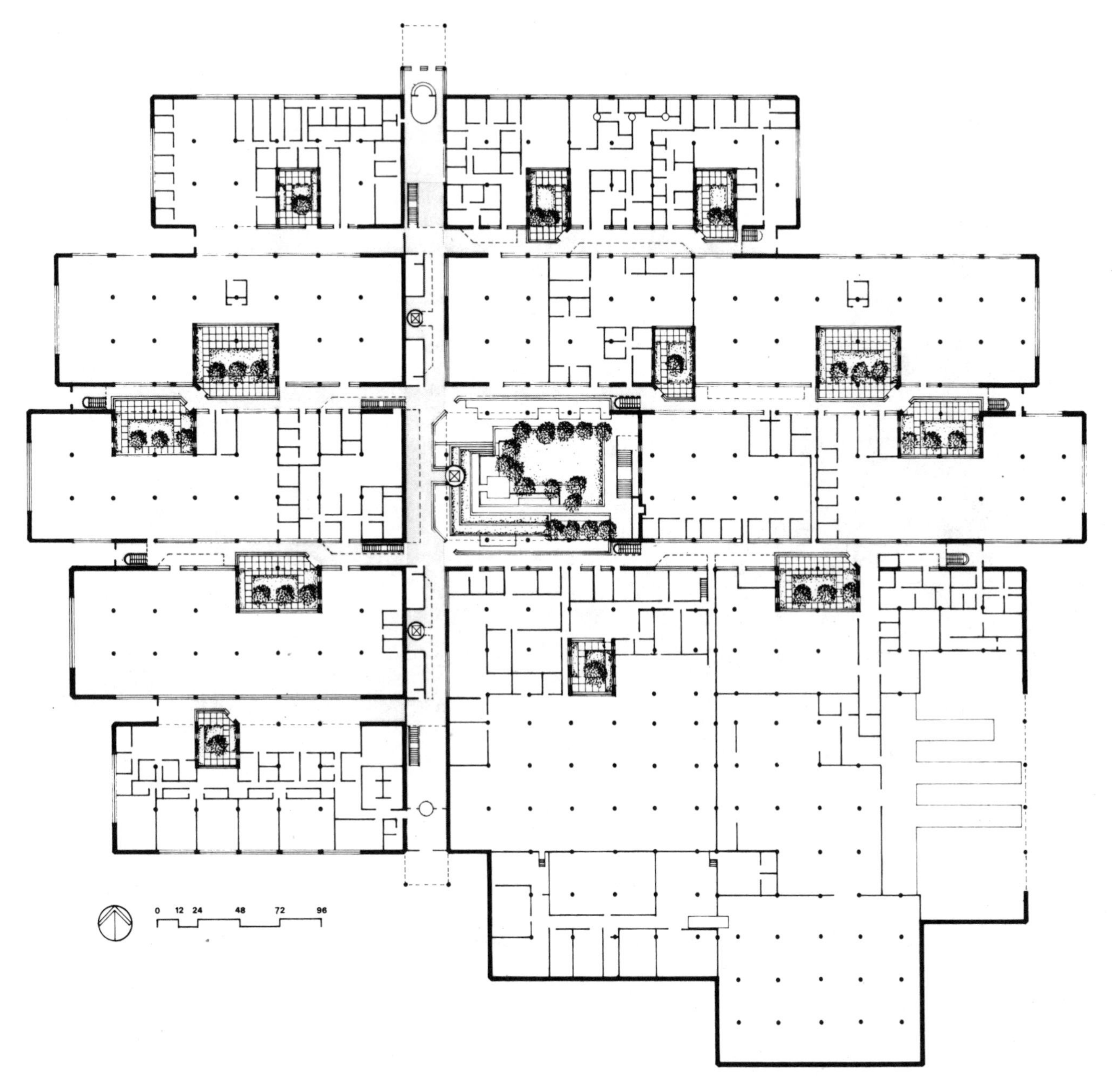

On ground floor, entrances at both ends of long hallway (shaded area) give access to all office areas and to parklike town center in middle of building.

Unlike conventional skyscrapers with identical floors stacked vertically and independent of each other, this $28 million Justice Department Building has a horizontal organization that eases work flow, improves user access, and takes advantage of the natural climate.

The 350,000-square-foot structure, built on an old fairgrounds in suburban Sacramento, was completed in 1982. It was the culmination of a five-year design/building program. This program involved extensive user participation throughout the design process. User group meetings were held and a

The exterior, of stucco and with simple lines.

newsletter was issued to appraise users of the project's progress.

The building houses facilities for criminal records processing, crime lab testing, and the instruction of police. In addition, the building includes administrative offices, a computer center, storage space for files, classrooms, luncheon facilities, and a print shop.

Because of the heavy programmatic demand for "paper moving," a horizontal organization was developed for the building. Circulation was a strong determinant of form; like a small city, the building elements are connected by a network of main and side streets, open-air courtyards, and a major central court. Lofty corridors occurring every 72 feet in the east-west direction, with clerestory windows, provide natural light for both floors. The corridors also act as giant ducts for cooling the building. Stairs and elevators act as vertical circulation elements. Between the two-story, toplit streets is a series of large blocks of space that serve as open office areas, small separate offices, and storage spaces for files. Natural light is provided to the office blocks through the streets and courtyards.

Harnessing the Energy of the Local Climate

The Sacramento area is blessed with a significant variation between daytime and night temperatures during the summer months. As the valley heats up during the day, the cooling breezes from the ocean are drawn through the Carquinez Strait and north toward Sacramento, where the temperature can drop from a high of 95° F. at 4 p.m. to a low of 58° F. at 6 a.m. During the ten to twelve hours when the temperature is below 70° F., this cool outside air can be used to lower the thermal mass of a building and thus decrease the air conditioning load during the rest of the day.

The Justice Building has a concrete frame structure, waffle slabs, and concrete shear walls exposed for maximum thermal mass. The two-story east-west "streets" have efficient, low-speed propeller fans at their ends, and serve as ducts to draw the night air through the building. Bypassing the static pressure of the duct work coils and filters, large volumes of night air are drawn directly into the ceiling plenum and through the corridors, flushing out the warm daytime air. This approach added $100,000 to the building cost, but will pay back in eleven years and save $245,000 in energy costs over twenty years.

Energy-efficient HVAC features include: recovery of computer center waste heat by double bundle condensers to heat air supplied to perimeter spaces; use of return air to extract heat from recessed fluorescent light fixtures; microprocessor controls; solar-supplied domestic hot water; variable air volume fans; and chilled water storage that takes advantage of lower nighttime utility rates.

In addition to 3½ inches of insulation on the exterior of the roof to enhance thermal storage effectiveness, walls are insulated to a U factor of .08. Windows are double-glazed. Reflective glass and minimum openings are used on the east and west facades to reduce solar heat gain. Almost all south glass on the southern facade is shaded from high summer sun by building recesses.

Through simple ventilation techniques, innovative lighting, and effective mechanical and control systems, the building is expected to use *38,000 Btu/sq. ft./year*. This is approximately half the state's required performance standards for such buildings of 75,000 Btu per square foot annually. Initial investment in mechanical, electrical, and solar systems is 29 percent of the building budget. Simple paybacks on the various features are estimated to range from three years for the computer-based building automation system, to twelve years for solar domestic hot water.

Corridors serve double purpose, for traffic and for air circulation.

Energy-Saving, High-Performance Lighting

Open-office layout and daylighting also reduce the energy requirement for lighting to about 1.5 watts per square foot per year. Widely spaced parabalume low-brightness fluorescent fixtures throughout the building provide a projected ambient lighting level of 20 to 30 footcandles or an alternate setting half that intensive. Supplemental fluorescent lighting on dimmers will provide 70 to 80 footcandles at work stations. Non-yellow high pressure sodium lighting in the corridors provides 15 footcandles. Low-voltage local switches and dual level ballasts are used throughout so ambient

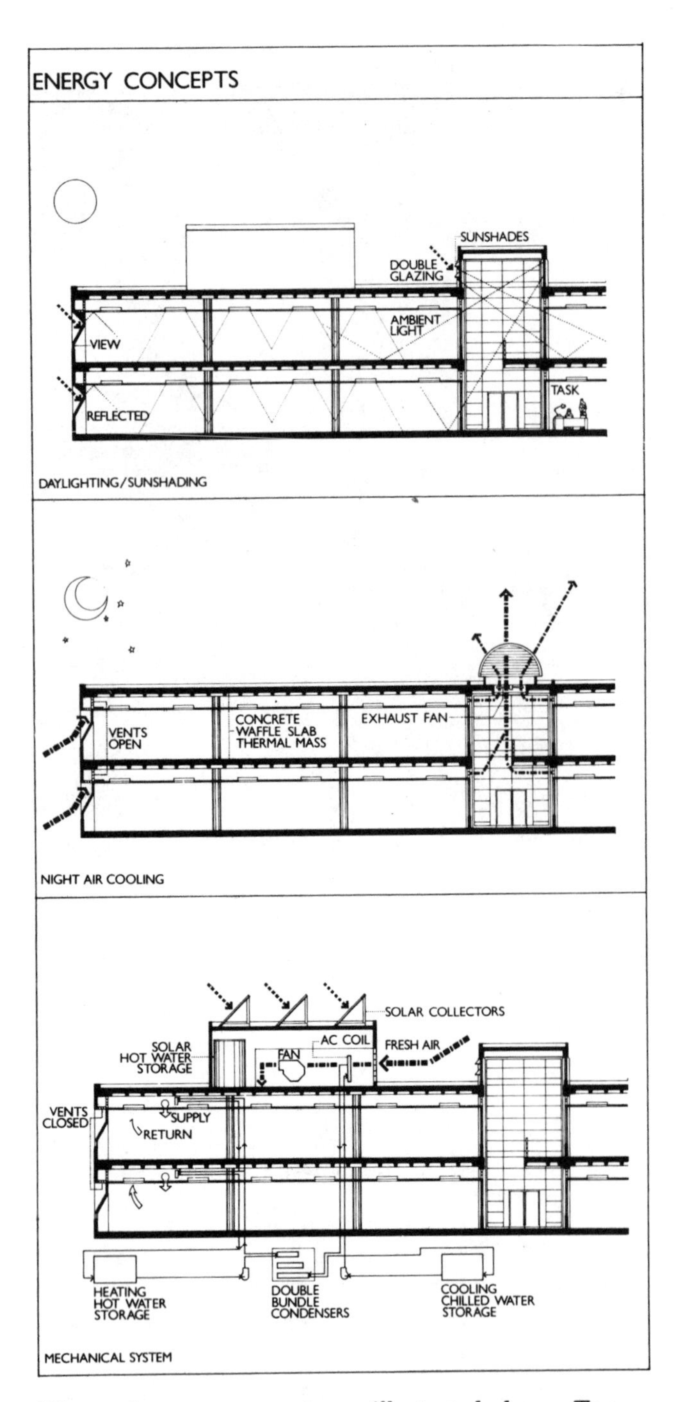

Three of energy concepts are illustrated above. Top, reflective glass, sunshades, and minimum openings reduce unwanted solar gain. Middle, for cooling, fans pull in cool night air, flush out warm daytime air. Bottom, mechanical system includes solar hot water system and chilled water storage.

and task lighting can be reduced or turned off when the space is not occupied, or when daylight is adequate.

Balancing Systems Choices for Energy Savings

Of the more than forty energy-saving features that characterize this building, the following are among the most important:

a. High thermal mass: cast-in-place concrete structure well insulated on the outside face.

b. Night air cooling: the end of each street is furnished with a low-velocity fan which flushes warm daytime air from the building with cool night air.

c. Sun-shading and daylighting: maximum use of shaded south light and north light; minimal east and west facing windows protected by reflective glass.

d. Low ambient and task lighting: local manual control of the artificial lighting environment, and photo cell control of fixtures near windows to benefit from daylighting.

Although some energy-saving features added initial cost to the building, a maximum payback of twelve years was determined for the most costly systems. Among these, the chilled water storage added $60,000 first cost to the budget. Its twenty-year energy saving is projected at $130,000. The solar hot water system, costing $26,000, is expected to produce a twenty-year saving of nearly $60,000.

Many energy-saving items added no cost to the project, yet will produce substantial dollar savings. For example, features in the building envelope to reduce heating and cooling loads will produce a twenty-year savings of $1 million. Others include: using reclaimed heat from lighting and computers; lowering of hot water temperature; variable temperature of chilled water to correspond with load requirements; reduced fan requirements (lower CFMs); and more efficient air conditioning components, such as variable air volume units and blow-through fans.

The twenty-year energy-system saving projected for the entire building is $4.4 million.

The Aesthetics of Security

Security control and ease of surveillance were resolved without creating the appearance of a fortress or an armed camp. The site planning and landscaping isolate the building from the surrounding residential neighborhood with a ring of earth berms and eucalyptus trees. The rather severe facades and bold earth red stucco are businesslike and unrevealing. There are no landscape elements near the building to hide the uninvited. Concern for security is best revealed in the employees' parking lot where the radial layout allows surveillance from the guard station just inside the doors. The regularity of the exterior has no nooks and crannies either horizontally or vertically to encourage snooping or sniping. Yet, the system of glazed corridors and office wings could be expanded outward in almost any direction. And the 16-foot-wide double-glazed roofs and clerestories of the interior streets and the accessible courtyards provide an airy spaciousness and high passive performance to a tight and densely organized plan within the secured building enclosure.

Windows have reflective glass to reduce solar gain.

A Pleasant Horizontal Skyscraper

Unlike skyscraper solutions for office needs that produce identical floors stacked vertically and independent of each other, this horizontal organization respects work flow, user access, and exposure to the natural climate in a more integrated three-dimensional plan. Both passive and hybrid means are used to conserve resources and to keep energy needs low. More important is the agreeable working atmosphere and the open arrangement that allow easy daily use as well as future flexibility within a secure envelope.

The colorful exterior stucco skin, well-designed landscape, and the solid construction insure a low maintenance project in addition to high performance.

The Department of Justice Building is an urbane and well-organized solution to a large building program. It provides a pleasurable and variable continuum of working environments for its many users and supports their complex and interrelated functions in an orderly but interesting horizontal design. It is an innovative resource-conserving, energy-conscious design, with clarity and light. It is a prototype of elegant energy architecture for working environments whose quality has been recognized by a number of national honors and awards.

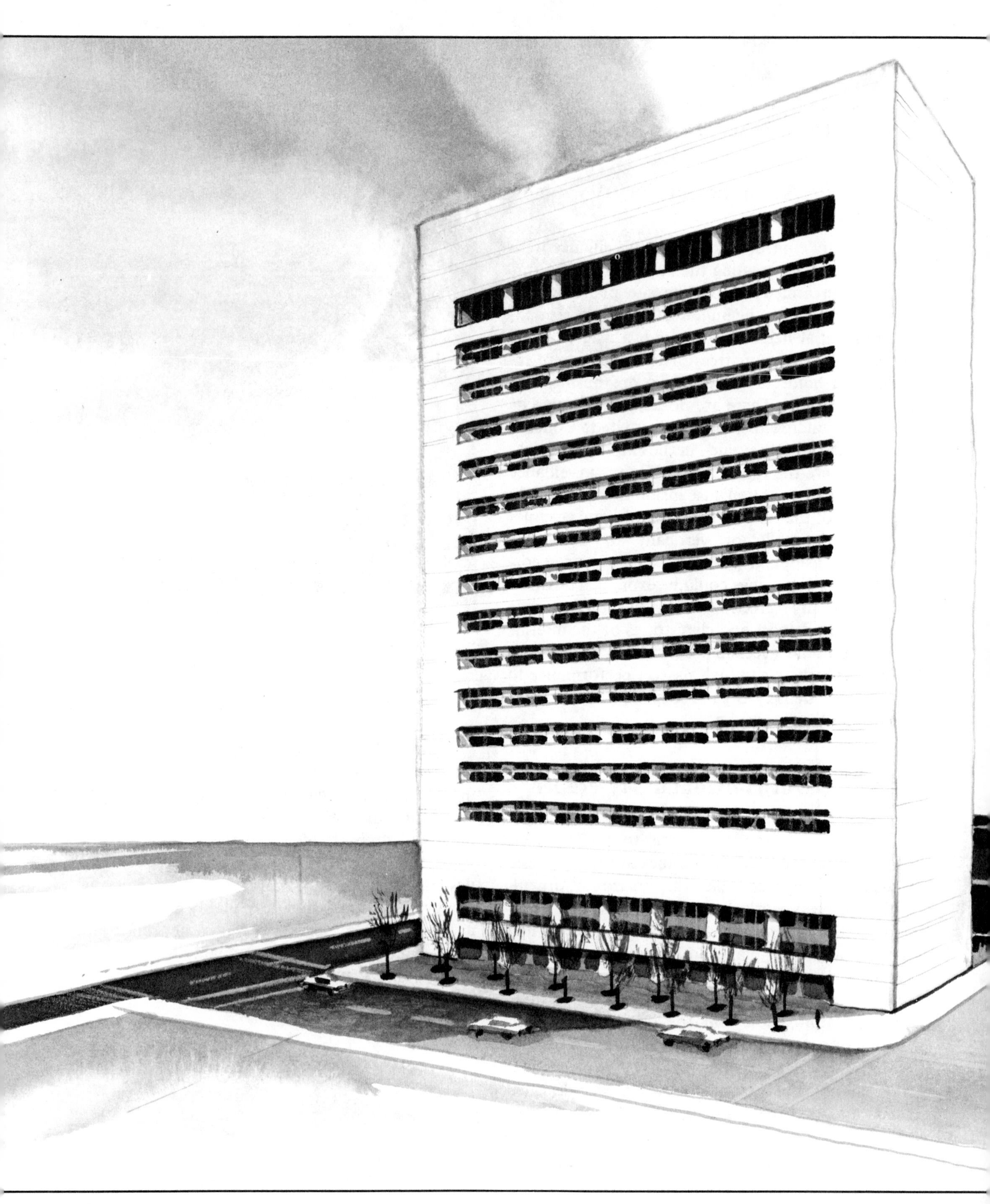

A Passive Prototype Office High-Rise

BUILDING

Farm Credit Banks of Spokane Office Building
First Avenue and Wall Street
Spokane, Washington

DESIGNER

Walker McGough Foltz Lyerla, P.S.
Architects and Engineers
Gerry Adkins, Director of Design
West 244 Main Avenue
Spokane, Washington 99201

STATUS

Occupied Fall 1982

CONTRIBUTORS

Vladimir Bazjanc, Energy Consultant
Box 4158
Berkeley, California 94704

Flack and Kurtz
Mechanical and Electrical Engineers
1425 Market Street
Denver, Colorado 80202

BUILDER

Hoffman Construction Company
900 SW Fifth Avenue
Portland, Oregon 97204

Estimated Building Energy Performance

BUILDING DATA

Heating degree days	6,650 DD/yr.
Cooling degree days	370 CDD/yr.
Volume of conditioned space	2,000,000 cubic ft.
Net area of conditioned space	225,000 sq. ft.
Building cost	$22,000,000
Cost of special energy features	$700,000
Total building UA	28,049 Btu/°F./hr.
Heating	
Net area of solar-heat collection glazings	14,000 sq. ft.
Energy required auxiliary from electric-driven heat recovery (Supplementary from gas fixed boilers)	4,265 Btu/ sq. ft./yr.
Cooling	
Energy required (electric) (Electric-driven refrigeration)	11,330 Btu/ sq. ft./yr.
Hot Water	
Electric energy required	94 × 10⁶ Btu/yr.
Gallons of water incoming at 45° F. delivered at 95° F.	226,000 gal.
Lighting	
Electric energy required	2.83 Kwh/ sq. ft./yr.
Connected lighting load (2,600 hrs./yr.)	334 Kw
HVAC Fans	
Energy required	1.85 Kwh/ sq. ft./yr.
Elevators	
Energy required	1.6 Kwh/ sq. ft./yr.

ANNUAL ENERGY SUMMARY

Nature's Contribution	
Heating	Cannot be isolated
Lighting	5,040 × 10^6 Btu.
Auxiliary Purchased Electricity	10,361 × 10^6 Btu.

a Lobby
b Interview
c Interview
d Word processing
e OJS offices
f Fire control
g Files
h Workroom
i Air shaft
j Loading dock

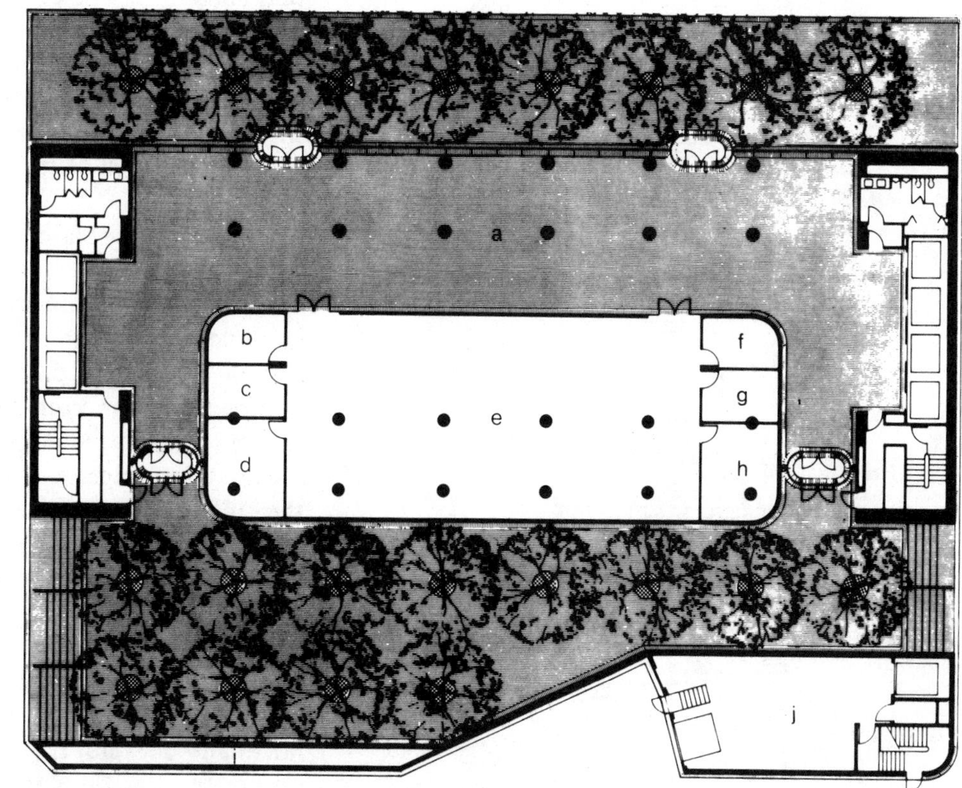

Ground level floor plan. Tree-sheltered entrances at top front on First Avenue, provide access to spacious lobby.

The Farm Credit Banks Building of Spokane, Washington, with its dramatic vertical slab, is an important prototype of the passive high-rise office building. It is among the first privately sponsored large buildings within the context of the new energy ethic that emerged in the 1970s. Occupied in 1982, it represents the creative collaboration of a strong local architectural firm with recognized national consultants in work for a progressive corporate client.

To call this 18-story building "solar" is only to identify symbolically its architectural mediation of the natural world. The major design strategy for saving energy was the use of natural light for the work spaces. This established an interior width of 65 feet as the most efficient dimension for cross lighting of open offices. But energy conservation, as parametically analyzed by computer modeling, was the key to the design process.

Design by Computer Analysis

A thorough energy analysis was undertaken prior to the start of the building design. It began with the investigation of various office building configurations and orientations and their impact on the energy consumption of a building in Spokane. The performances of rectangular, square, and triangular building shapes were simulated and compared. Each architectural shape and orientation was analyzed to determine its energy use profile. The building shapes were also compared for their daylighting performance, since this was identified early as an energy-effective design strategy. The architectural massing forms compared were analyzed with clear, tinted, and reflective glazing as well as with and without a complete shading sys-

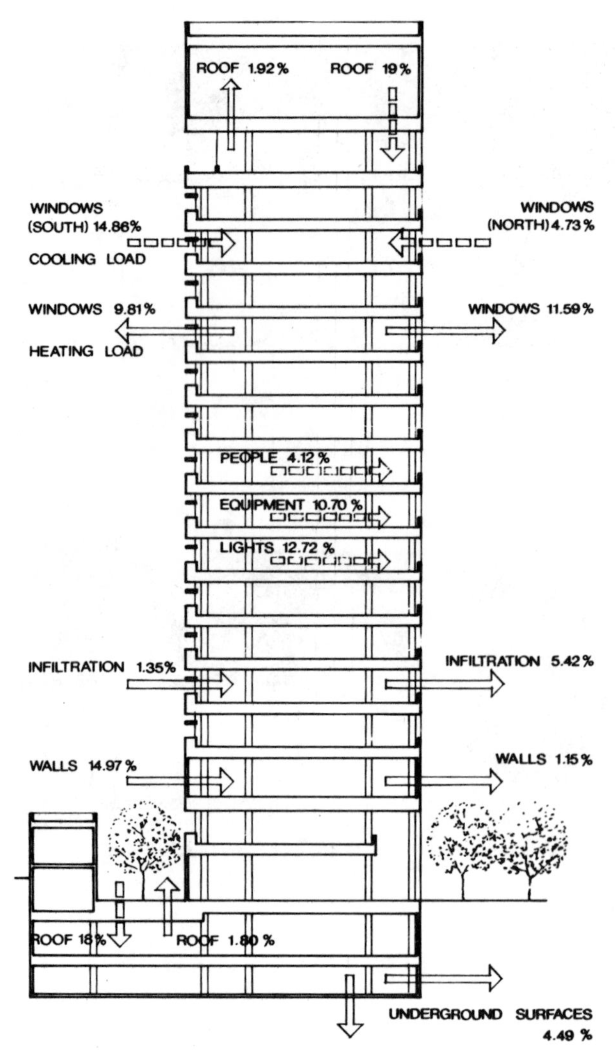

The heating and cooling loads, with heating loads indicated by solid arrows, cooling by divided arrows.

tem. Consistently, a rectangular building shape facing north-south proved to be the most energy-conservative building. Fortunately, the best views coincided with the north-south orientation for windows.

Designing the Window Wall

Computer studies also showed that complete shading of glass could reduce cooling loads in the building by more than 30 percent. This confirmed that shading is one of the most effective strategies of energy conservation in the design of such a large building for office uses. The shading system consists of exterior horizontal and vertical surfaces positioned in front of glazing on the south facade. The outside edge of these surfaces is 3 feet in front of the glazing.

Horizontal shading surfaces are 6 feet 4 inches and 9 feet above each office floor. Both horizontal shading surfaces and the window sill reflect daylight deeper into the interior. This total shading system completely prevents exposure of windows to direct sunlight from April until September. During winter months, it allows limited direct sun exposure to supplement the heating requirements of the building. Due to the low sun angle in the morning and late afternoon, solar heat gain renders the east and west facades virtually impossible to deal with effectively in terms of environmental comfort. Thus for energy reasons, the building's service cores were located on the east and west ends of the building. The north facade is oriented directly north and does not require sun shading in this location in the Pacific Northwest.

Extensive analysis was performed to explore possible alternatives in glazing assembly. The results indicated that shaded double pane windows with clear glass offer the best performance considering both heat and light on an annual basis. Glare is controlled by the upper sun shelf and interior roller-mounted vinyl solar shades that can be operated manually by occupants.

Responding to the Urban Context

There is a refined sense of inevitability about the completed building. But the open site and the enlightened client allowed the full variety of possible configurations to be explored by computer simulation. The site is a compact square plot across the end of a block and thus facing on three streets. Almost any architectural configuration was possible on the 27,000-square-foot site within Spokane's central business district.

An east-west plan dimension of 178 feet was established by using the full dimension of the site along First Avenue; it offers the greatest north-south exposure for offices and affords the best views. This long east-west dimension also allowed for the maximum use of natural light, thereby reducing the amount of artificial lighting. The exterior dimension was fixed at 69 feet with the 4-foot difference to the inside being devoted to solar shading of the exterior south facade. The 65-foot dimension allows for the deep penetration of natural light into the offices from both sides on a year-round basis and minimizes middle zones that cannot be lighted naturally.

The tower is positioned midway on the site, allowing for the development of an exterior plaza on both the north and south sides. The north face of the building, set back from the building line along First Avenue, visually widens the street. The brick-paved plaza with two rows of trees provides an urbane forecourt to the spacious building lobby with the north facade towering in a wall 18 stories high.

Below ground level there are two levels: the sub-basement and basement levels are devoted to support functions, such as storage, mechanical spaces, print shop, and training center. A public lobby and support office spaces occupy the ground and mezzanine levels. The third floor houses a complete computer center, and floors 4 through 16 provide office space for the bank and rental space for tenants. The 17th floor, the cafeteria, includes an exterior dining terrace on the south side of the building, a place to enjoy the best views. The typical configuration of each office floor, totally open, with the service cores containing elevators and toilets at both ends, allows maximum flexibility in office layout, and in effect turns inside out the traditional core plan of a high-rise building.

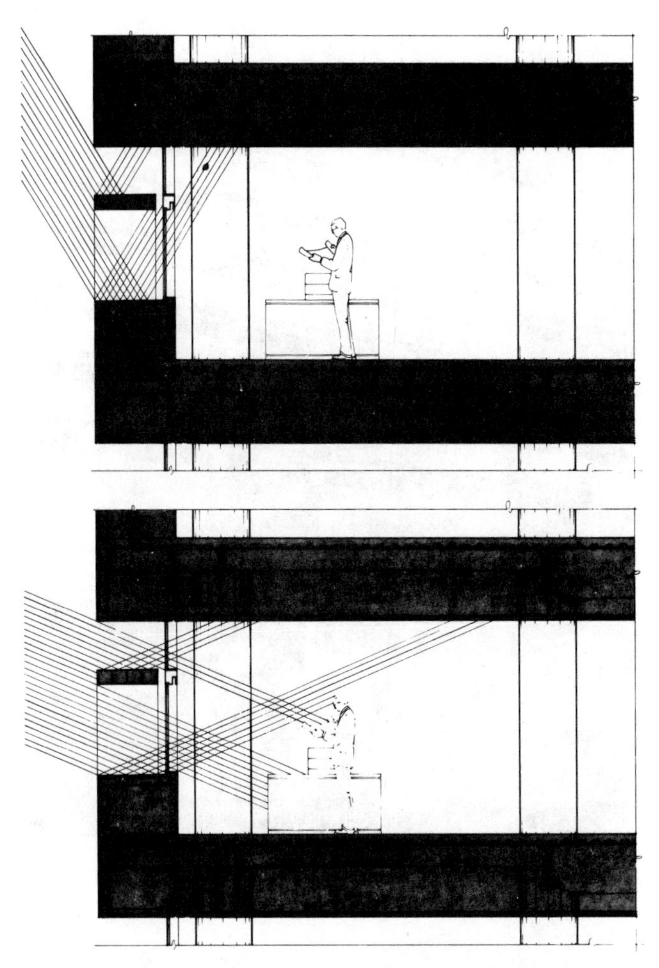

Above, patterns of sun at noon on August 21, when only small amount of sunlight can penetrate the building. At noon on December 21, sunlight and warmth are reflected deep into the building.

Economics in the Mechanical and Electrical Systems

The mechanical and electrical systems include a number of novel features and techniques that result in a minimal yearly energy consumption of 40,000 Btu or more per square foot per year.

Innovative design and detailing have resulted in a high-performance building that is well under the energy consumption guidelines established by ASHRAE 90-75 and by various federal government agencies such as the Department of Energy and the General Services Administration.

Average annual lighting energy consumption of 1.4 watts per square foot is achieved through the use of newly designed fixtures that produce high quality light distribution and effective illumination without glare. Standard parabalume overhead fluorescent lighting fixtures controlled by photocell dimmers respond to daylighting conditions. Lighting levels in task areas were projected at 71 footcandles; corridors and lobbies at 30 footcandles.

The pattern of windows soaring eighteen stories.

For calculation purposes the average electrical demand was reduced by 20 percent to allow for the daylighting contribution based on the automatic dimming control.

To select the optimum power plant system, the architects compared 14 alternative heating and cooling plant types, using the Ross F. Meriwether Energy System Analysis Series and Department of Energy DOE-2 computer programs. Energy consumption estimates included heating and cooling, exhaust, lighting, electric power needs, and service water heating.

Economic comparisons covered amortized initial capital investment and costs for energy, maintenance, operation, taxes, and insurance on all the systems on an annual basis for 20 consecutive years.

The energy requirements were derived from a computer model simulation program that integrated hourly weather and solar data with internal load estimates, equipment shut-off, and setback time schedules, setback temperature schedules, and percentage variation profiles for all internal, process, electric power, lighting, and airflow loads, and central heating and cooling equipment.

The system with the best energy result included electric heat recovery chiller, 150,000-gallon thermal storage, "free-cooling" cycle, and gas boiler. Fan motor energy for circulating conditioned air throughout the building is reduced by the use of variable air volume (VAV) supply systems, augmented by fan-powered induction boxes. The thermostatically controlled boxes made use of heat from lights in the ceiling, using only the minimum amount of air supply necessary to provide individual comfort in the various spaces. Efficient after-hours use of the office tower is achieved by having dampers in the supply connection at each floor, which only open to admit air as required for that floor.

The mechanical system uses heat recovery (heat pump) refrigeration machines to transfer internal heat to the perimeter of the building at an efficient energy rate, enhanced by the use of water tanks for insulated thermal storage. A free-cooling cycle is provided with water from the cooling tower circulated directly in the chilled water system, to permit cooling service without the operation of the refrigeration compressors when climatic conditions permit. A computer management system automatically indexes fan equipment and heating and cooling equipment to supply only as much energy as is needed for comfort in the occupied spaces.

A hydronic system interconnects various pieces of plant equipment to maximize their function and achieve optimum system operation for the least energy usage. The various functions are achieved by the automatic control of equipment and water flow through valves to minimize the use of pumping.

Achieving a Prototype

In comparison to the apparent architectural simplicity of the building form, the mechanical and electrical systems seem to involve decisions of great complexity. In fact, all levels of design decision-making were informed through that sophisticated tool of artificial intelligence, computer simulation. The key was the team of human intelligence who knew what the correct questions were, and knew when it had an answer.

Aside from the building itself, there are several aspects of the design that should be noted. Although many individual designers participated in a decisive and responsible way, no single great name can be identified to credit this building. Similarly, the client is also a team, a responsible corporation obviously with a special heart and soul, but not dominated by a single leader.

The skyscraper is often credited as a major American contribution to the world of architecture. As a building type, it is a product of speculatively high-cost land and the uniformity of office practice as expressions of cultural need. The skyscraper was invented in the most advanced commercial center of its time, Chicago, and has been involved in similar urban concentrations such as New York. But this humane variant of the skyscraper that almost diagrammatically epitomizes the solar energy ethic has been conceived and built in a small handsome city on the opposite coast. This neat passive high-rise office building has the added advantage that it is easy to understand.

The prototypical aspects of the Farm Credit Banks Building have been honored by a number of awards and publications besides the First National Passive Awards program. The simplicity and clarity of the building form demonstrate that energy response does not require intricacy or complexity. But it does require depth, both in the investigative analyses of the team at every stage in the design process, as well as an architectural depth revealed, for instance, in a study of the geometric resolution of passive energies in the walls. It took new design tools as well as a new disposition toward energy use to generate this new prototype of the hundred-year-old skyscraper form. Although superficially the techniques of daylighting or shading may seem to be the theme of this office high-rise, in fact they are only components of a comprehensive design discipline that infused every design decision with a new spirit.

Appendices

Appendix A

SUNDANCE ONE FOR VIRGINIA

Passive solar features of Sundance One, the Roberts' home in Reston, Virginia, clearly saved energy for the owners. The building heat load was met by 52 percent solar energy, 37 percent internal gains, 7 percent electric baseboard heat and 4 percent fireplace heat. The solar system saved 17.38 million Btu or 5,089 kwh of electrical energy. This represents a savings of $356. These figures were established by an extensive monitoring program, from February through July 1982.

The building cooling load was 24.80 million Btu. The building lost 3.57 million Btu via conduction, the auxiliary heat pumps removed 0.07 million Btu and passive solar cooling removed the remaining 21.16 million Btu. The passive cooling fraction was 99 percent, since only a small quantity of auxiliary cooling was required. The cooling system used a total of 2.41 million Btu of operating energy for the fans and the thermal curtain. The net solar cooling savings were 18.75 million Btu. This equals 5,490 kwh of electrical energy or $384, at $0.07 per kwh.

The total savings for heating and cooling were $740 for February through July 1982 for an average of $123 per month!

The automated movable insulation controls the quantity of solar gains in summer and heat losses in winter from the 663-square-foot Trombe wall. The large internal thermal mass (264,000 pounds) acts to store solar energy for heating, and also as a buffering mechanism to reduce summer temperature swings. The building was quite comfortable during both the heating and cooling seasons measured, while using little auxiliary energy. By using automated movable insulation, the building provides excellent levels of performance. The building is a good example of an integrated passive solar heating system which produces no penalty during the cooling season. During July 1982, the occasional use of the heat pumps was necessary to remove accumulated water vapor from the building more than to reduce building air temperatures. This is typical of mid summer operation. The total heating and cooling system is well suited for its climate zone according to the monitored data.

Space Heating Analysis

During the test period, the Trombe wall and sunspace systems had a total of 58.33 million Btu of solar radiation available. A total of 19.00 million Btu was collected. The Trombe wall collection efficiency was 32 percent, while the sunspace collection efficiency was 33 percent. The movable insulation subsystem required 0.50 million Btu to operate. Thus, the solar collection Coefficient of Performance (COP) was 38.0. A total of 17.88 million Btu of solar energy was used to serve the 34.16 million Btu building heat load. When internal gains were subtracted from the building heat load, the equipment heat load totalled 21.45 million Btu. Auxiliary heating was provided by 2.73 million Btu of electric baseboard heat and 0.85 million Btu of fireplace energy. The occupants burned about one-third of a cord of hardwood. The storage system increased in energy content by 1.12 million Btu by admitting 18.57 million Btu and releasing 17.45 million Btu to space heating. Ninety-eight percent of the solar energy to the load was delivered via storage. Direct gains were small with the sunspace contributing about 6.58 million Btu of the collected solar energy. Most of the sunspace solar energy heated the upper level of the home, while 1.89 million Btu were transferred by fans into the rear mass air-core wall.

The automated movable-insulation system reduced heat losses through the single-glazed Trombe wall. The overall computed U value of the operational Trombe wall system was 0.11 according to the data. The system controls opened the thermal curtain at 9:30 a.m., and closed it again at 4:30 p.m. on a typical, sunny winter day.

Space Cooling Analyses

During the test period, the performance of the passive solar cooling subsystem was excellent. Auxiliary cooling energy was small, 0.07 million Btu, and was used only at the end of July. The thermal curtain closed when the outdoor temperature exceeded 72° to 75°F. This provided for avoidance of direct solar gains to the Trombe wall, which averaged just 75°F. in temperature during May through July. The efficiency of the collection subsystems was reduced intentionally to 15 percent for the Trombe wall and 20 percent overall. The metal decking was rolled down over the sunspace glazing to reduce its transmission to 0.25. Only 5.91 million Btu of solar energy were collected, of the 30.08 million Btu available. Total losses from the collection subsystems were 24.17 million Btu. The internal gains, including fan operating energy, totalled 7.13 million Btu. Of this, 3.93 million Btu were absorbed into the storage. Thus, a total of 9.84 million Btu was transferred to the storage masses. The storage temperature increased from 70° to 76°F. during May through July. Storage energy content increased by 0.71 million Btu during this period. Of the 9.13 million Btu released from storage, 5.20 million Btu increased the cooling load. The storage mass worked very well, reducing temperature swings during the cooling system. A total of 21.16 million Btu of thermal energy was removed from the home by a combination of the building's resistance to net radiant gains and through night venting via windows, cool tubes, and the twin thermal chimneys. The thermal chimneys have been shown to remove approximately 23,300 Btu per hour at an 8°F. difference in temperature from inside to the outdoors.

Appendix B

T.V.A. SOLAR MODULAR HOUSING

These are the standard specifications for the T.V.A. solar modular housing.

Floor

1. Rim joist framing double 2 × 12s.
2. Floor truss, 24 inches on center.
3. ¾-inch T&G underlayment grade plywood decking, glued and fastened.
4. R-19 fiberglass insulation between joists with .040 asphalt impregnated, reinforced Kraft paper applied to bottom of joists.
5. Vinyl linoleum floor covering in kitchen, baths and utility room area.
6. FHA/VA carpet with ½-inch pad in living room, hall, dining and bedroom.

Walls and Partitions

1. 2 × 6 framing stud grade, 24 inches o.c. with corner bracing.
2. All openings are double framed with appropriate-sized headers to insure maximum strength.
3. Double top plate; sound methods of solid framing are practiced.
4. ½-inch finished gypsum.

5. 6-mill poly vapor-barrier and R-19 fiberglass insulation.

6. ¾-inch insulated sheathing applied to all exterior walls.

7. Pre-finished siding in choice of color and style.

8. Exterior electrical outlets gasketed to vapor barrier.

9. Sill plate caulked.

Roof/Ceiling

1. Wood truss system with steel pinch-plate gussets, 24 inches o.c., sheathed with ½-inch CDX plywood.

2. 235-pound, self-seal asphalt shingles installed over equivalent 30-pound felt.

3. ½-inch finished gypsum ceiling with a 6-mill poly vapor-barrier.

4. R-38 fiberglass insulation installed between roof truss.

5. 96-inch ceiling heights.

6. Aluminum soffit system with ridge vents and/or roof louvers.

7. Eaves front and rear with 24- and 12-inch rakes.

Windows

1. Casement, Thermopane—insulated glazing; fully weather-stripped, prefinished interior; vinyl wrapped.

2. Operable panels equipped with screen and safety latch.

3. Window quilts all windows.

Doors

1. Wood-framed, exterior steel doors; insulated, weatherstripped, complete with lock and dead bolt; dimensions, 3 feet by 6 feet, 8 inches.

2. Pre-finished interior doors, in sizes appropriate for the rooms, are 6 feet, 8 inches high and 1⅜ inches thick.

3. All jambs, trim and moldings are prefinished to match.

4. White metal, bi-fold closet doors.

Cabinets

1. Factory-built wood cabinet—both kitchen and bath.

2. Color-coordinated, one-piece laminated-plastic countertop (self-edged).

3. Four-inch high back-splash (as applicable); reversed side of top sealed.

4. Mirror over vanities.

5. Side-mounted medicine cabinet.

6. ⅝-inch painted or prefinished particle board shelving in wardrobes and linen closets.

Plumbing

1. All systems according to the National Plumbing Code.

2. Vitreous china water closet and tank.

3. 5-foot fiberglass tub/shower combination with overflow and equipped with tub and shower fixture.

4. Enamel-steel lavatory equipped with single-lever faucet, pop-up drain and overflow.

5. All water-distribution systems are CPVC with shutoff valves to all fixtures.

6. Stainless-steel, double-bowl kitchen sink with single-lever faucet.

7. All drainage and vent system is ABS Schedule-40 plastic.

8. Plumb for washer.

9. Shower rod.

Electrical

1. All electrical wiring according to the National Electrical Code.

2. 200-amp, 30-branch panel.

3. All lighting fixtures U.L. approved (standard lighting package).

4. Special outlets: range, dryer, refrigerator.

5. Ceiling light fixtures in kitchen, bath, hall and bedrooms.

6. Hanging light over dining and breakfast area.

7. Switched receptacle in living room and dining area.

8. Vanity light over lavatory.

9. Exterior front and rear wall-mounted light fixture with interior switch.

10. Ground-fault circuit interrupter in bathroom and exterior.

11. Smoke detector.

Heating/Cooling

1. Forced-air system or perimeter design with metal insulated duct.

2. All floor registers in living room, dining room, and bedrooms are adjustable.

3. Wall-mounted adjustable registers in bath, kitchen and utility rooms.

4. All-electric heating systems (heat pump optional).

5. Bath vents in baths per plans.

Molding

1. All trim moldings are factory pre-finished to match doors.

2. 2¼-inch prefinished door molding.

3. 2¼-inch prefinished base molding.

Decor

1. Surface-mounted towel bar and tissue holders in bath.

Appliances

1. 52-gallon gas water heater used in calculations.

2. Other appliances optional or per customer order including Rheem Solar DHW system.

Hardware

1. Entry doors equipped with two keys (not master keyed).

2. All applicable door hardware mortised.

3. Privacy lock sets in bedrooms and bath.

4. Passage hardware all other interior doors.

Appendix C

SOLAR CALCULATIONS FOR ELLIOT MULREADY HOUSE

Typically, solar consultants work within set climatic zones and building types, and develop a judgment about how to interpret raw data. Michael Riordon is a well known California solar consultant, especially familiar with the variable climates of the bay area. Although he has a strong scientific background with a Ph. D. in physics, he is known in the national solar community as the co-author of *The Solar Home Book* and as editor of Cheshire Books. In the following discussion, Riordon's familiarity with the quirks of local climate and the realities of domestic solar applications is reflected in the series of judgments in his short, complete method of establishing average annual solar performance.

Solar Heating Estimates

The design goal in sizing the greenhouse/rock storage bed system for this home was to be able to store *half* the solar heat trapped inside the greenhouse on a sunny January day. The sizing method used was developed by the solar research group at Los Alamos National Laboratory under Douglas Balcomb and published as *The Passive Solar Design Handbook, Vol. 2.*

First, the solar heat trapped inside the greenhouse was estimated from ASHRAE Solar Heat Gain Tables for 40° north latitude to be about 400,000 Btu on a clear January 21. This number does *not* include the sunlight reflected off or absorbed in the glass—more than 100,000 Btu.

Next, assuming a thermostat inside the greenhouse activates a fan or blower at 85 °F., we would expect the rock bed to cycle from 65 °F. to 75 °F.—a temperature change of 10 °F. From the Los Alamos method, the following system characteristics are needed to store 200,000 Btu, or 50 percent of the total heat trapped on January 21:

— 48 tons 2-inch rock (1,000 cubic feet)
— 1,500 cubic foot per minute blower or fan
— 20 square inch minimum duct cross-sectional area

In this configuration, we could expect a maximum heat transfer of 33,000 Btu per hour from greenhouse to rock bed.

This is a *very large* storage system, but it is possible. Almost 40 percent of the solar energy striking the greenhouse glass is eventually stored as heat in the rock bed. Such a percentage may only be possible in very mild climates like California, where outdoor air temperatures are 55 to 65 °F. on a sunny January day.

Of the remaining 50 percent solar heat trapped inside the greenhouse, about 30 percent is lost back through the glass; and about 20 percent, or 80,000 Btu, is transferred into the house by passive means—mostly warm air flowing into the house by natural convection. Another 85,000 Btu of direct solar heat gain is received through the southeast and southwest window glass, for a total of 365,000 Btu solar heat gain on a clear January day.

Of course, not all January days are sunny. One way to adjust for this fact, when making monthly heat loss and heat gain estimates, is to multiply the sunny-day figures by the number of days in the month and by the percentage of possible sunshine for that month, which is about 55 percent in the San Francisco bay area in January.

This is a rough but effective estimate:

365,000 Btu/day × 31 days × 0.55 = 6.28 Btu × 10^6 where 10^6 Btu = 1 million Btu

From independent heat loss calculations, the Elliot/Mulready house design loses heat at the rate of 9.72 Btu per day per square foot of floor space per °F. temperature difference between the indoor and outdoor air. As the total heated floor area is 2,383 square feet and there are 518 heating degree days, locally on the average in January, the expected heat loss is:

$$9.72\left(\frac{\text{Btu}}{°\text{F. ft}^2\ \text{day}}\right)2383\ \text{ft}^2 \times 518\ °\text{F. day} = 12.0 \times 10^6\ \text{Btu}$$

Therefore the percent solar heating in January is

$$\%\ \text{solar} = \frac{6.28\ 10^6\ \text{Btu}}{12.00\ \text{Btu}} = 52.3\ \text{percent}$$

and 5.72 × 10^6 Btu, or 47.7 percent, must be supplied by auxiliary heating systems.

The figures of 50 percent solar heat stored in the rock bed, 20 percent transferred to the house by passive means and 30 percent lost back through the glass are accurate enough for coldest months—December, January and February. But as outdoor air temperatures rise, the house simply can't use all the heat and some or all of it will have to be dumped—either by venting the greenhouse or by not running the fan/blower. In all other months, the percentage of heat transferred to the rock bed or house will be lower than the December–February figures. The percentage of possible sunshine data also changes from month to month, as do the degree days for that locale. Repeating the January calculations for the eight other months in the heating season, we arrive at the following table of performance estimates.

Summary

SOLAR GAIN PER SYSTEM

Month	*Hdd*	*% Sunshine*	*Heat Loss (× 10⁶ Btu)*	*Heat in Rockbed × 10⁶ Btu*	*Passive Heat from Greenhouse into House × 10⁶*	*Direct Gain × 10⁶*	*Total Solar Heat Used (× 10⁶ Btu)*	*Aux. Heat (Mbtu)*	*% Solar Heating*
JAN	518	.55	12.0	3.41	1.36	1.51	6.28	5.72	52.3
FEB	386	.60	8.9	3.5	1.4	1.5	6.4	2.5	72
MARCH	372	.65	8.62	3.87	1.55	1.77	7.19	1.43	83
APRIL	291	.70	6.74	2.8	1.2	1.2	5.2	1.02	77
MAY	210	.70	4.86	2.1	.90	1.47	4.47	.33	92
JUNE	120	.70	2.8	1.5	.4	.90	.28	0	100
OCT	137	.70	3.2	1.8	.5	.90	3.2	0	100
NOV	291	.60	6.7	2.8	1.1	1.51	5.41	1.3	81
DEC	474	.50	11.0	3.0	1.2	1.3	5.5	5.5	50
TOTALS	2799	—	64.82	24.79	9.61	12.06	46.45	17.8	72%

The Elliot/Mulready house annually requires a total of about 65 million Btu to keep it warm and gets about 72 percent of that, or 46 million Btu, from the sun. The monthly solar heating fraction ranges from a low of 50 percent in December to 100 percent in June and October.